IMAGES
of America

HOUSTON FIRE DEPARTMENT

On the Cover: Members of Hook & Ladder Company No. 1 pose with their ladder wagon in the late 19th century. On August 14, 1838, the city fathers formed Protection Fire Company No. 1 and constructed a firehouse for the company at the corner of Fannin and Preston Streets. Its motto was *Semper Paratus* ("Always Prepared"). Its first piece of equipment was a large barrel on wheels with handles. (Courtesy of the Fire Museum of Houston.)

IMAGES
of America

Houston Fire Department

Fire Museum of Houston
and Tristan Smith

ARCADIA
PUBLISHING

ISBN 978-1-4671-3298-5

Published by Arcadia Publishing
Charleston, South Carolina

Printed in the United States of America

Library of Congress Control Number: 2014954330

For all general information, please contact Arcadia Publishing:
Telephone 843-853-2070
Fax 843-853-0044
E-mail sales@arcadiapublishing.com
For customer service and orders:
Toll-Free 1-888-313-2665

Visit us on the Internet at www.arcadiapublishing.com

To all firefighters—past, present, and future—who have served and given their lives to protect us.

To Jen, whose love and support has made all the difference.

CONTENTS

ACKNOWLEDGMENTS

Since the Fire Museum of Houston opened in 1982, hundreds of staff and volunteers have poured their blood, sweat, and tears into preserving the history of Houston's fire service. It would be impossible to thank every single person individually, but, without their hard work, their love for Houston's history, and interest in the fire department, this book would not have been possible. The Fire Museum of Houston's research library, archives, and photograph collections span the history of the city's fire service and beyond, telling compelling stories and information about the growth of Houston, not just the city's fire service. The staff members and volunteers who maintain these collections cannot be thanked enough. Unless otherwise noted, all images appearing in this book were provided by the Fire Museum of Houston.

I also wish to thank my editor at Arcadia Publishing, Stacia Bannerman, for her expert guidance, support, and prodding. There are several people who have paved the way for this book through their research, writing, and enthusiasm for Houston's fire history. Specifically, I wish to thank Tom McDonald, Emily Ponte, Angela Rayne, Ken Paradowski, Max McRae, Dan Kopfensteiner, Dave Hamm, Jackie Haynes, Herb Kimbrel, Daniel Kopfensteiner, Tim Kopfensteiner, T.J. Zalar, James Godlove, Ray Boyer, Cheryl Byington, Ron Conn, Bill Edge, Jay Evans, Thomas Greider, Mary Lou Henry, Rick Flanagan, Diane Kopfensteiner, David Massarano, David Thompson, Jim and Cheryl McGee, Calvin Mendel, Scott Mellott, Glen Mears, Mike Pack, Jerald Ricks, David White, Joe Russo, Tom Hopkins, Dave Miller, Paul Box, Jeffrey Caynon, Chuck Buschardt, Joe Gay, and Duane Troxel for their many contributions toward preserving our fire community's history.

Finally, to Jen, thank you for your love and encouragement, not just on this project, but in everything. And to Theo, Max, and Antoinette . . . thanks for putting up with me, supporting me, and cheering me on while I have pursued my dreams and passion over the years.

INTRODUCTION

Only two years after the Allen brothers, Augustus and John Kirby, put their townsite up for sale in 1836, several Houston businessmen organized Protection Fire Company No. 1. The bucket brigade was designed to protect their investments from fire. By 1859, two other brigades had organized, and the three groups joined to form Houston's first volunteer fire department.

Following the Civil War and Reconstruction, Houston's growth in population and size called for an expanded fire service. The original three brigades were joined by six more companies, and a new device was enlisted in battling fires: a steam fire engine. The engines required fewer men to operate and pumped hundreds of gallons of water per minute. They immediately proved more powerful and efficient than the hand-operated pumpers.

As the 19th century came to a close, the city's fire companies had outgrown their efficiency as volunteer organizations. All opportunities for improving the efficiency of the volunteer department were explored before it was decided that a paid city service was needed. The rapid growth of the city, the increase of financial support the city was giving the volunteer companies, and a series of devastating fires forced immediate change. In 1895, Houston transformed the volunteer companies into the Houston Fire Department (HFD). The transition went smoothly. Thomas Ravell was selected chief of the new department, and he selected 44 of the volunteers to staff the department. Ravell used seven of the former volunteer fire stations, and the city purchased most of the volunteers' apparatus and horses.

Between 1900 and 1910, the department doubled its manpower, several new stations were constructed, and new companies were organized in the outer neighborhoods. In 1913, an annexation doubled the city's land area and included the port's turning basin. It was during this period that the department began transitioning from horse-drawn apparatus and began modifying its vehicles with internal combustion engines. By 1921, all vehicles had been motorized, replaced, or attached to motor tractors and towed to fires.

With the city having transitioned to a paid fire service, the department grew in leaps and bounds. The motorized fire apparatus enabled the department to do more work as it grew, but with fewer men at fewer new stations. Additionally, the scheduling constraints that the first firefighters were accustomed to began to ease up. A second shift was added, and firemen were able to alternate workweek day and night shifts on a monthly basis. This made it easier for the men to spend time with their families every day. In order to further protect his firefighters, Fire Commissioner Allie Anderson—a former firefighter—was able to open the city's first fire-training facility, adjacent to the Central Station, in late 1928. The facility cost $22,000 and included a five-story tower and a miniature building used as a laboratory for ventilation classes.

However, many within the department were at the mercy of politics. The outcome of mayoral and commissioner elections in Houston during this time usually forecast any major changes in the fire department's administration. Houston's last fire commissioner and former city councilman, Frank Mann, recalled that firings, demotions, and new fire chiefs were commonplace when new commissioners took office. Firemen were governed by city-controlled civil service rules, until those rules were placed into the hands of the Texas legislature in 1947. By 1945, the citizens of Houston had seen the fire department's trucks motorized, firefighters working a double-platoon system, the purchase of the first fireboat, construction of a new Central Station, the establishment of the first training facility, and the building of an independent fire-alarm office.

Houston grew nearly fivefold in the 1950s. Annexations on two occasions more than doubled the city's size in each case. In the first, the department assigned reserve apparatus to the former volunteer stations, with three becoming permanent HFD facilities. Communities served by 13 other volunteer stations were taken into the latter, with seven kept as permanent. Volunteers were expected to help fight fires initially, as additional Houston fire units were dispatched from stations closer to town.

Aside from the motorization of apparatus, Houston's firefighting methods went mostly unchanged for the first half of the 20th century. However, three technological improvements impacted Houston's firefighting: fog nozzles, self-contained breathing apparatus (SCBA), and the one-and-a-half-inch hose. The hose enabled firefighters to better battle blazes in single-family homes, the nozzle gave them a variety of water-spray options, and the SCBA allowed firefighters to enter buildings with less regard to the smoke being produced by burning material.

At the beginning of 1963, a third shift was created, reducing the average workweek. Firefighters now worked in three-day intervals consisting of 10-hour days and 14-hour nights, and then they were off three full days. Over 100 promotions resulted in order to handle the new shift. In late 1965, the city council raised firefighter salaries to spur recruitment. Top pay for veteran firefighters was the lowest for paid firefighters among comparable cities. The low pay hampered recruitment, and only three men frequently manned pumper companies. Staffing of ladder companies sometimes fell to two laddermen.

The department's facilities saw rapid growth and improvements in the late 1960s and early 1970s. A new four-story headquarters opened in early 1968. Fire Station 1 occupied the first and second floors, administrative offices were on the third floor, and offices of the fire prevention and arson departments occupied the fourth floor. In April 1969, a new central motor-repair shop and maintenance complex went into service, a new fire-training facility opened later that year just south of Hobby Airport, and a new fire alarm building followed in 1975 across the street from the new headquarters.

The 1970s brought economic prosperity for Houston. During the 1970s and early 1980s, the city, with a heavy oil-based economy, prospered as the price for crude oil rose to record levels. However, by the mid-1980s, the city was in the midst of one of its worst economic slumps ever, as crude prices dropped dramatically. It took nearly a decade for Houston's economy to recover as city leaders worked to diversify the economic base. The Houston Fire Department's growth during those years mirrored that of the city. Between 1978 and 1983, over 1,000 firefighters were hired; fewer than 300 were hired in the following five years.

The 1970s and 1980s also saw a rapid increase in diversity throughout the department. The city's Hispanic and Asian populations grew rapidly, and, in an attempt to reflect this demographic shift, minority hiring in the department increased. In 1992, Fire Marshal Eddie Corral was named fire chief, becoming the first Hispanic chief in the city's history. The number of active women firefighters in the department also grew. The first female firefighter was hired in 1975, but seven years passed before the second woman was hired. Recruitment of women into the department began to increase sharply in 2000.

Changes also came in services provided by HFD. District Chief L.O. "Whitey" Martin was chosen to head up a new emergency medical service (EMS) in the early 1970s after the city council placed pre-hospital emergency care within the fire department. Chief Martin selected 126 firefighters to become the department's first emergency medical technicians (EMT). With additional training, they became its first paramedics. HFD also increased its responsibilities at both William P. Hobby and Houston's International (now George H.W. Bush) Airports.

The generous city budgets of the 1970s and early 1980s disappeared by the mid-1980s, thanks to the regional depression that took place in the states with oil-dependent economies. The fire department began to face new challenges in the face of slim budget growth. In the mid-1980s, a number of major changes got under way after a management study of the department was released and the city faced lower budget projections. Management was decentralized, a variety of divisions saw a transition to civilian personnel, the EMS delivery system became a detailed incident-command procedure, the new 911 telephone system was implemented, six of the twenty-two fire districts were eliminated, and four quadrants were established. Deputy chiefs were moved out of headquarters to command the new quadrants, and they reported directly to the fire chief. In addition, the photography section and street index were closed down. Arson investigators now had to take their own photographs. The aides were transferred to fire apparatus; chiefs would have to drive themselves. If a chief needed an aide at a fire, they grabbed one of the firefighters on the scene. The largest acquisition of apparatus in the department's history took place in this year: 22 engines, 16 ladders, and 30 ambulances were purchased. In 1987, mobile automatic status terminals (MAST) were installed on all fire and EMS units. The terminals ended the excessive radio transmissions on the fire frequency. A button on the MAST was pressed by the officer to report his status, which was recorded in the dispatch computer.

Calls for help to the HFD have increased dramatically over the past decade. Much of the increase has been related to EMS incidents. However, because the department's EMS utilizes fire companies as first responders, the call volume for companies has increased dramatically as well.

In 1995, Houston expanded its city limits again to stretch out to the 1960 area and Kingwood area. This resulted in HFD building a temporary Station 96 at West Greens and Mills Roads to help service the Willobrook Mall area, which was completed in 2000. The Kingwood annexation resulted in the Kingwood Volunteer Fire Department disbanding after HFD took over or built new stations.

On May 25, 1991, the department added a fourth shift, which created hundreds of new promotions. Under the new work schedule, firefighters had to work one of their off-days about every six weeks to keep the average workweek at 46.7 hours. District Chief Lester Tyra replaced Chief Corral in 1998 after a new mayor took office. Chief Tyra was the fifth fire chief who had served previously as president of the firefighters' union (1970–1990). Chief Tyra immediately instituted a self-evaluation of the department, resulting in a five-year strategic plan, risk analysis of fire districts, and recognition as a leader department in 2000 by the country's fire chiefs and city managers. Tyra also won approval for the purchase of fifty-five engines, six ladder trucks, three ladder towers, and a hazmat foam engine. (The new equipment was to replace 60 percent of the first-line apparatus.) In 1999, thermal imagers, which improved the search for victims and hidden fire, were bought and placed on all ladder trucks, hazmat units, and rescue trucks. Fire Prevention was beefed up with the addition of 17 new fire inspector positions.

On February 23, 2010, the investigation of the April 12 fire of the previous year concluded that a variety of mistakes had been made that resulted in the deaths of a fire captain and a rookie. Among the mistakes were the separation of firefighters in the structure and radio procedures not being followed. No disciplinary action was taken.

On January 28, 2011, plans were announced to refurbish former fire station 27 in Denver Harbor into apartments for firefighters coming to Houston for medical treatment. Former Houston city councilman Peter Brown put up an initial $53,000 for the property and looked for an additional $400,000 to complete the project. Brown's father had designed the station in 1940. The Hamilton Brown House was the first of its kind in the country.

Due to the ongoing Texas drought, on September 13, 2011, Houston firefighters were dispatched to the park responding to a wildfire. The fire had quickly grown, and the firefighters were sent to the north levee near Interstate 10 to wait for the blaze to come to them and stop it there. By the end of the first day, over 50 vehicles and 175 firefighters had been dispatched to the scene. At one point, the fire was one mile wide, consuming 1,500 acres of the park, with the cause under investigation. On May 31, 2013, Houston suffered its largest casualty in a single event, when four firefighters were killed and thirteen others injured while fighting a five-alarm fire at Southwest Inn in southwest Houston.

As the population, size, demographics, and landscape of Houston grow and change, the fire department nips at its heels. In addition to the City of Houston's fire service, numerous other fire departments, many of them volunteer, preserve the suburbs (like Katy and Sugar Land) and island cities (such as West University Place or Southside Place). What does not change is the fact that Houston's firefighters are there to serve the city and protect its citizens and their property. Houstonians can sleep safe at night, knowing that the Houston Fire Department has their back.

One

THE EARLY YEARS

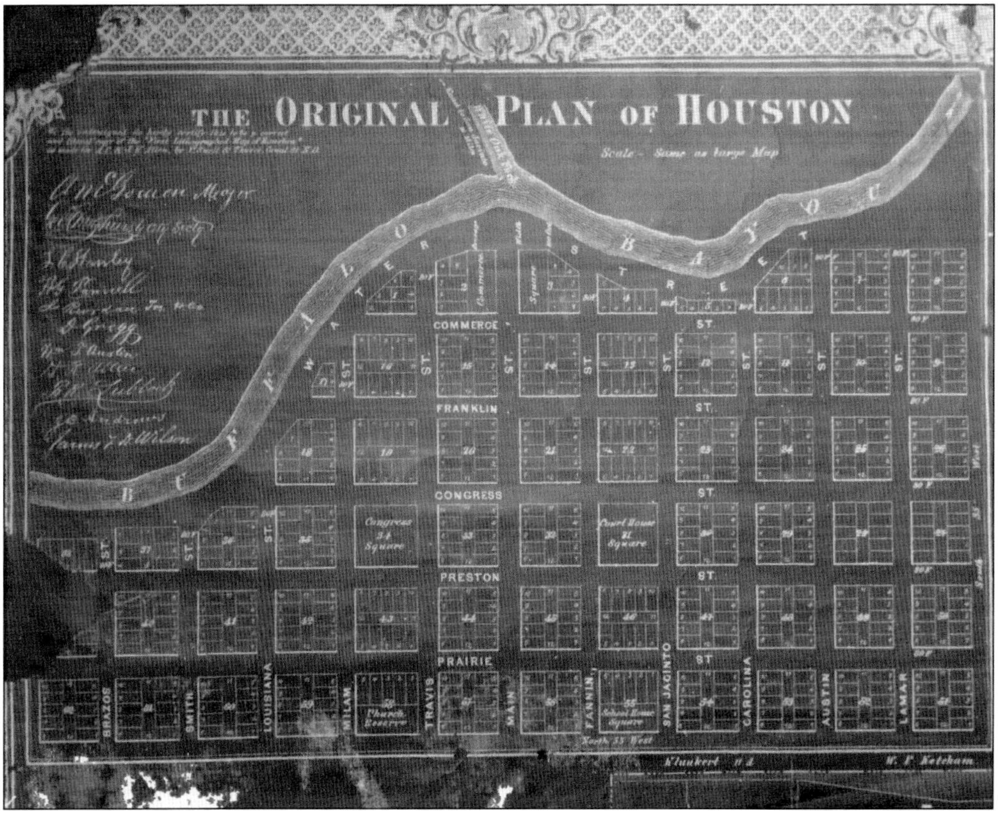

Surveyor Gail Borden drew the city's original plan for brothers Augustus and John Kirby Allen. The town contained 62 square blocks on the south side of Buffalo Bayou. Blocks 31 and 34 along Congress Avenue were dedicated to the construction of a county courthouse and a capitol building for the republic. On August 14, 1838, Houston's first volunteer firefighting company was established, Protection Fire Company No. 1.

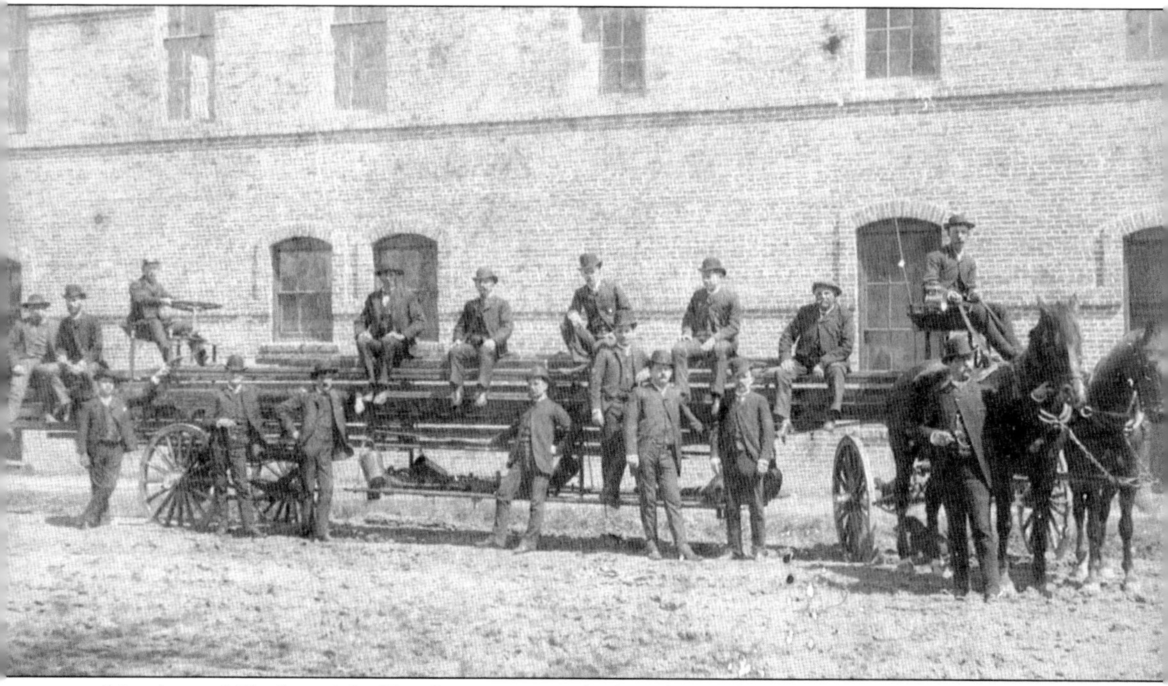

Members of Hook and Ladder Co. 1 pose with their horse-drawn ladder wagon in 1890. This company, comprised of wealthy Houstonians, formed on April 17, 1858, and received its charter the following year. They fancied themselves the elite among the three fire companies in town and housed their truck on Congress Avenue opposite Market Square. They disbanded in June 1895, as the department became a paid institution.

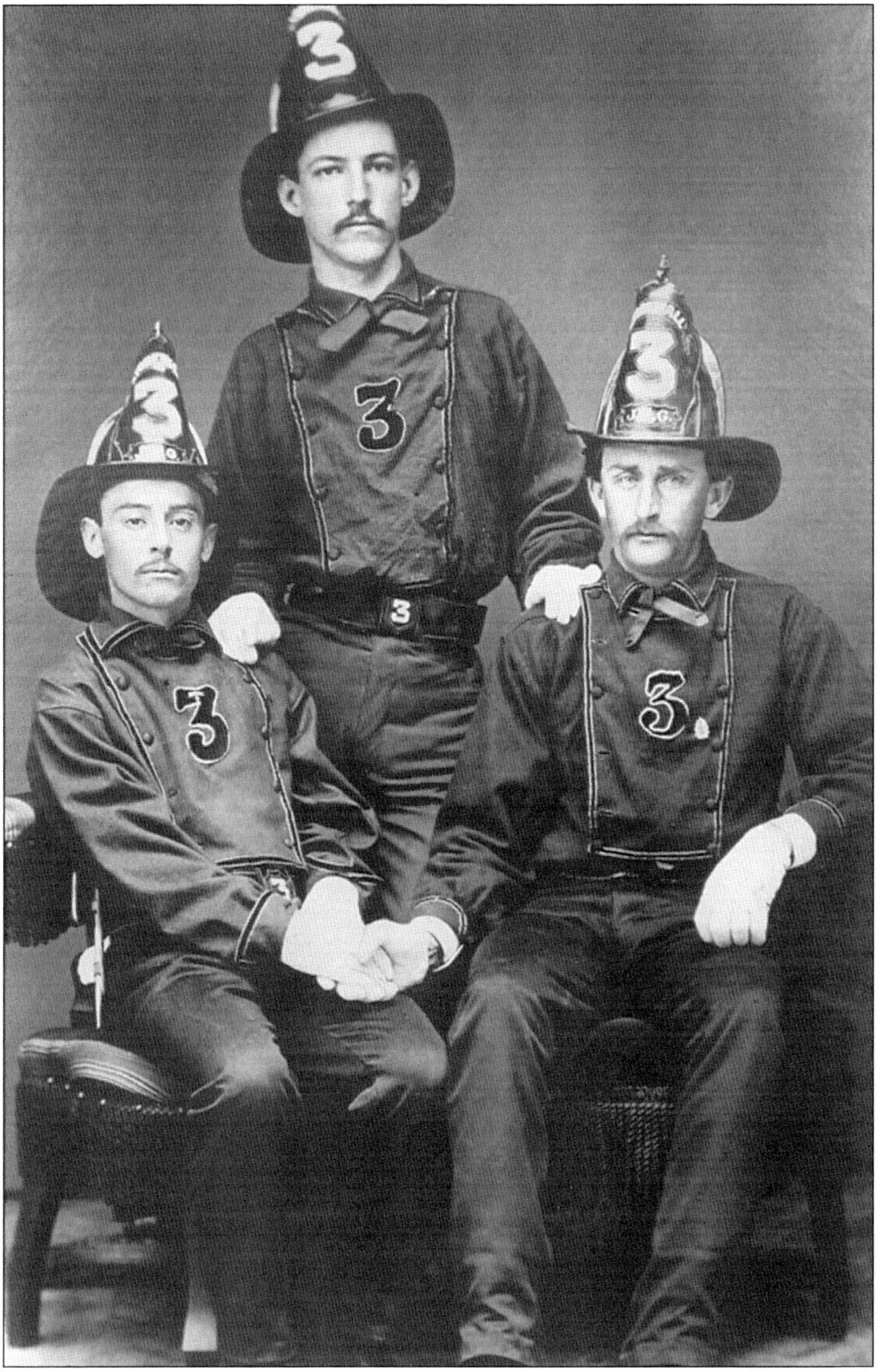

Here, members of Volunteer Company Stonewall Hose Company No. 3 pose for a photograph in the late 19th century. Stonewall 3, organized in 1868, had as its company motto, "Trust in Us." The company was composed of men who had returned from the Civil War under the command of Stonewall Jackson. The dress uniforms shown here were rarely worn to a fire but were used in parades and at social events.

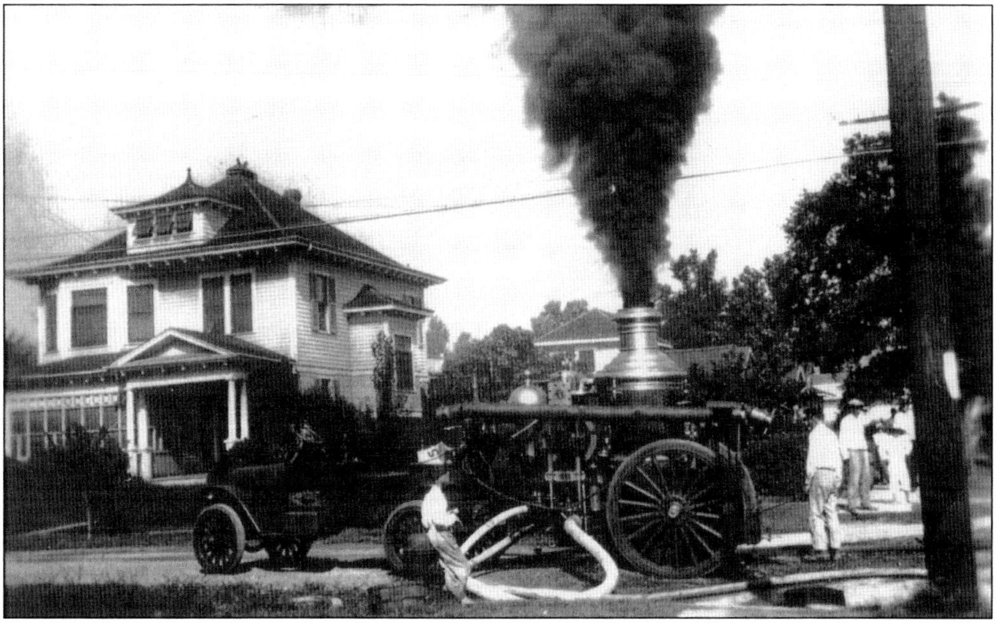

Houston's first steam-powered pumping engine was delivered in the late 1850s. In 1877, two more were purchased, but these were later repossessed. An additional one was delivered in 1894. By 1925, all the first-line engines were motorized. All five of the reserve engines were still steam powered—three with motors, and two pulled by tractors. By 1930, only one remained.

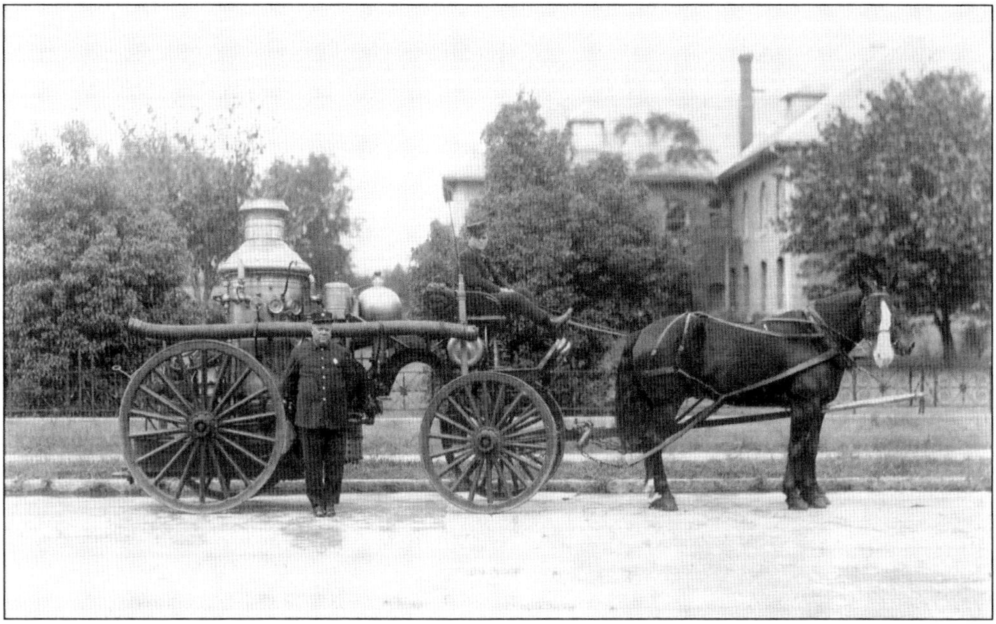

Just after 2:00 a.m. on October 16, 1894, a fire was discovered in the boardinghouse adjacent to Saint Joseph's Infirmary. Water pressure was weak, and the fire spread to the infirmary. The nuns were aided by firemen, police, and citizens in evacuating the patients. In the process, two nuns died and one was severely burned. Before daylight, nine of the twelve buildings on the block were destroyed. One of the major apparatus used in fighting this fire was the steam powered pumper, which was quickly becoming outdated as a major tool in fighting fires.

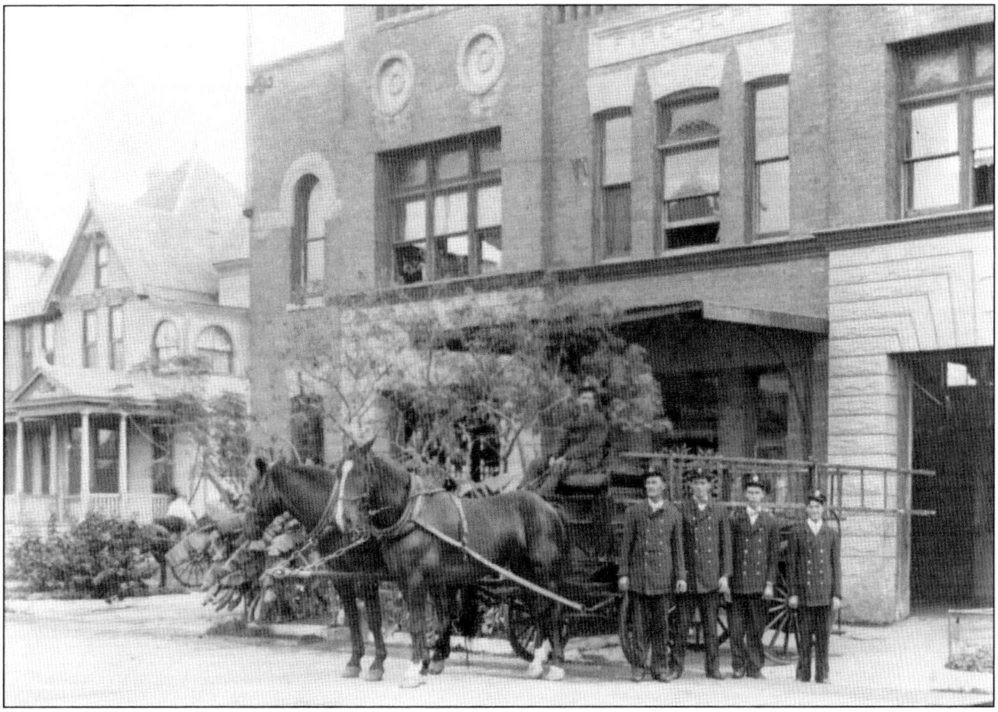

Station 7 opened in January 1899 as the city's first paid fire department station. It served Houston for 70 years before closing in 1969, when the crew moved to a new facility on Elgin Street. Following its preservation in 1981, the station opened as a museum. It is listed in the National Register of Historic Places, is a Texas Historic Landmark, and is a protected Houston landmark.

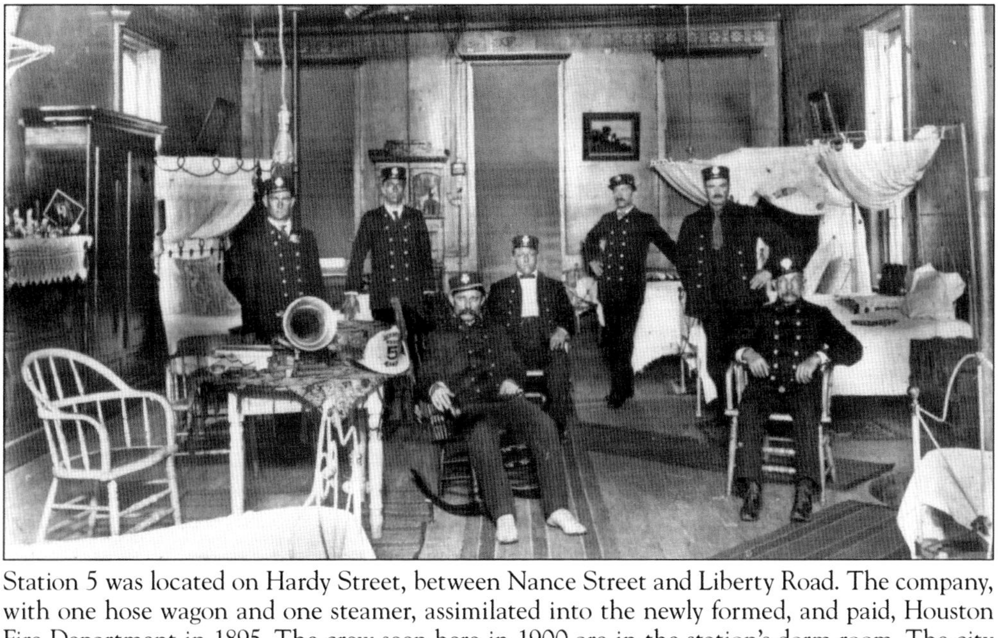

Station 5 was located on Hardy Street, between Nance Street and Liberty Road. The company, with one hose wagon and one steamer, assimilated into the newly formed, and paid, Houston Fire Department in 1895. The crew seen here in 1900 are in the station's dorm room. The city would begin replacing these volunteer stations in 1899 with the opening of Station 7 on Milam and McIlhenny Streets.

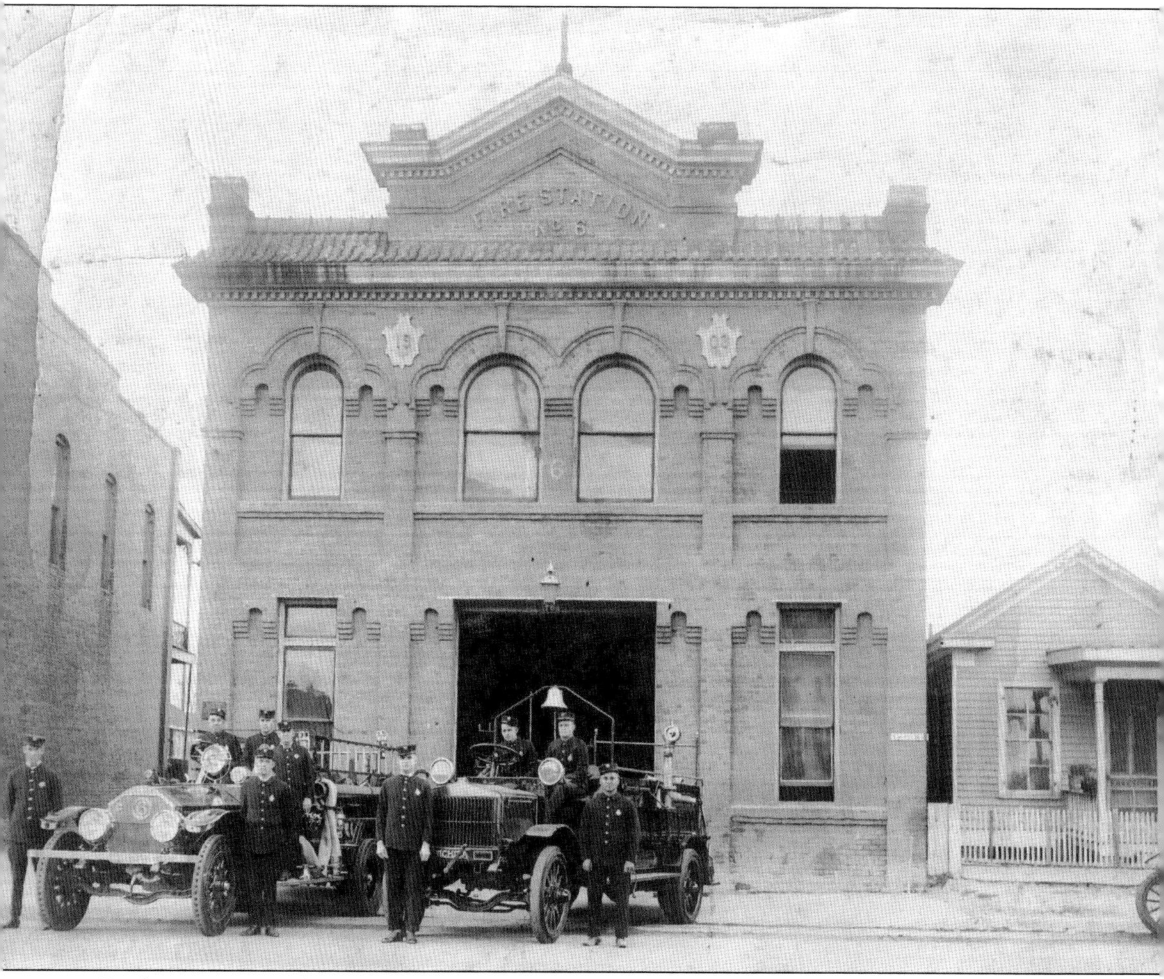

Firehouse 6 opened for service on Washington Avenue in 1903. In 1931, a new station was constructed for the crew on Henderson Street. The original station served a variety of uses until 2005, when the building, then in serious disrepair, was purchased by the owner of a local marketing company. It was restored to its original condition and repurposed for use as company headquarters in 2012.

The department's first paid fire chief, Thomas Ravell (right), and his driver, Louis Bradley, are seen at Central Station on San Jacinto and Caroline Streets. During the 57 years prior to becoming a city department, Houston's fire service saw twenty-four fire companies, only nine of which were members of the Houston Volunteer Fire Department, and at least eight of which were junior companies, composed mostly of preteen boys.

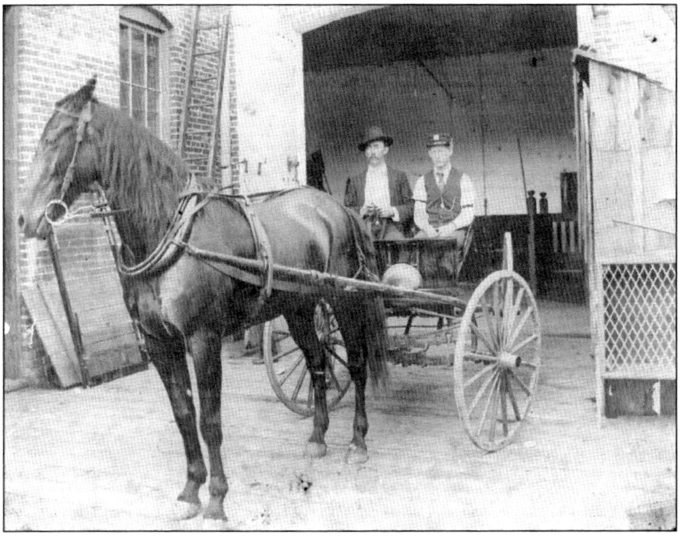

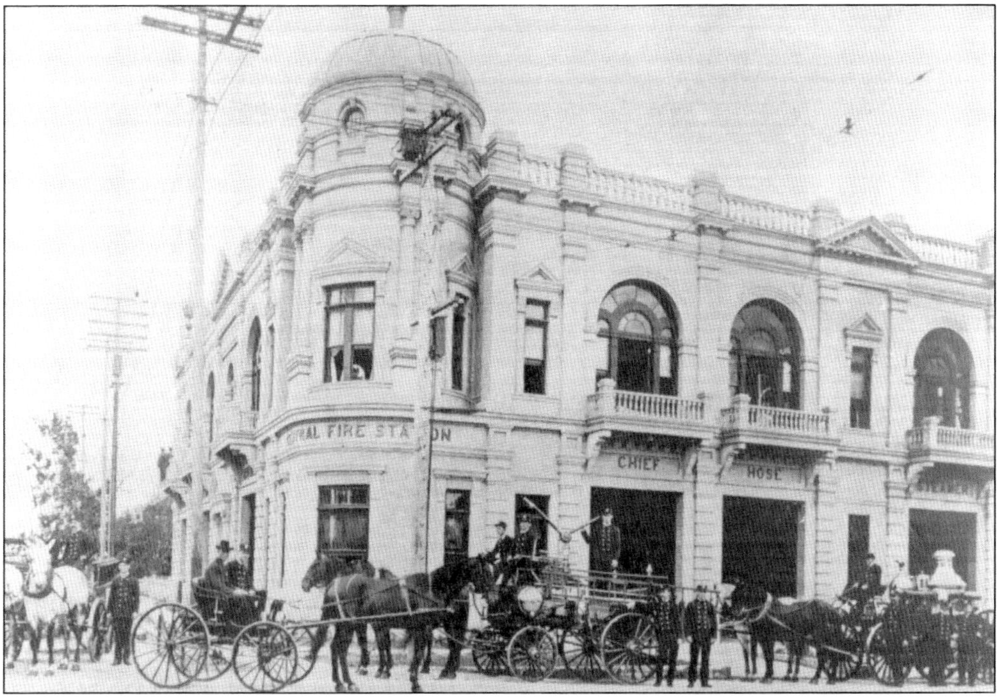

Central Fire Station was the first fire department headquarters building constructed and owned by the city. It replaced the building taken over by the paid city department, constructed in 1880. It burned down in 1903. The new station, constructed in 1904 at 519 San Jacinto Street, was home to the fire dispatch office, which was responsible for taking calls and monitoring the city's 50 fire alarm boxes.

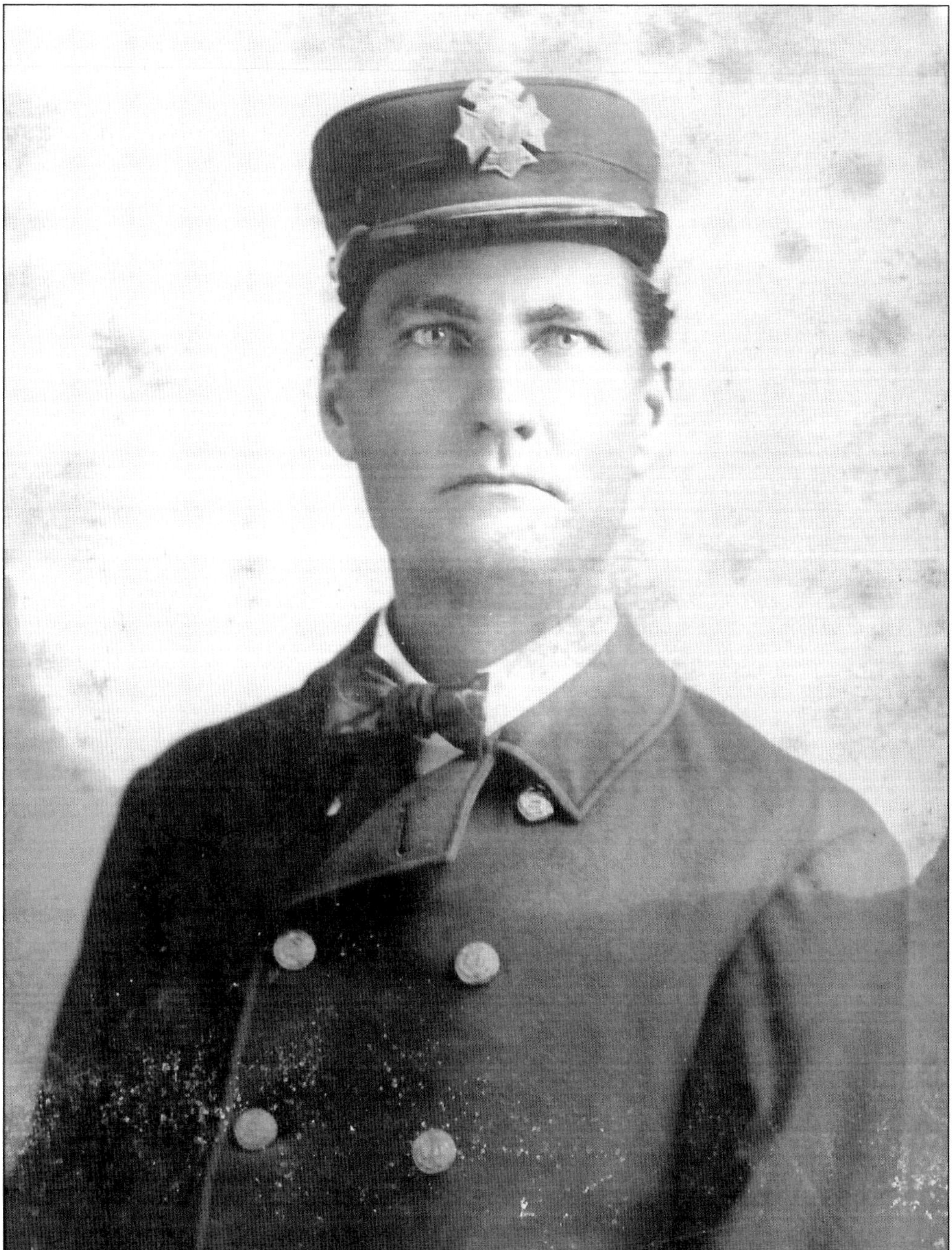

Capt. Drew Campbell King joined the Houston Fire Department in 1898. He served the city for almost 40 years, obtaining the rank of captain of Firehouse 7. He was instrumental in the formation of the department's first union. King's strong disciplinarian will was backed by the Colt .45 he was known to carry. He ended his shifts with a bucket of beer.

The building pictured here, located on the city's Market Square, was Houston's fourth city hall, replacing a building destroyed by fire in 1901. The new building, larger but similar in design to the previous one, opened in 1904. When the current city hall was opened in 1939, the 1904 building was converted into a bus terminal. It burned down in May 1960.

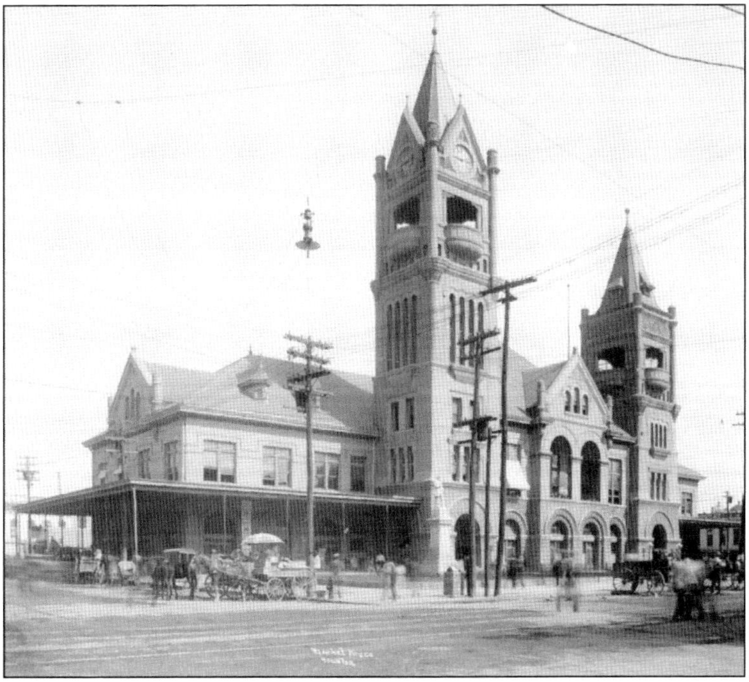

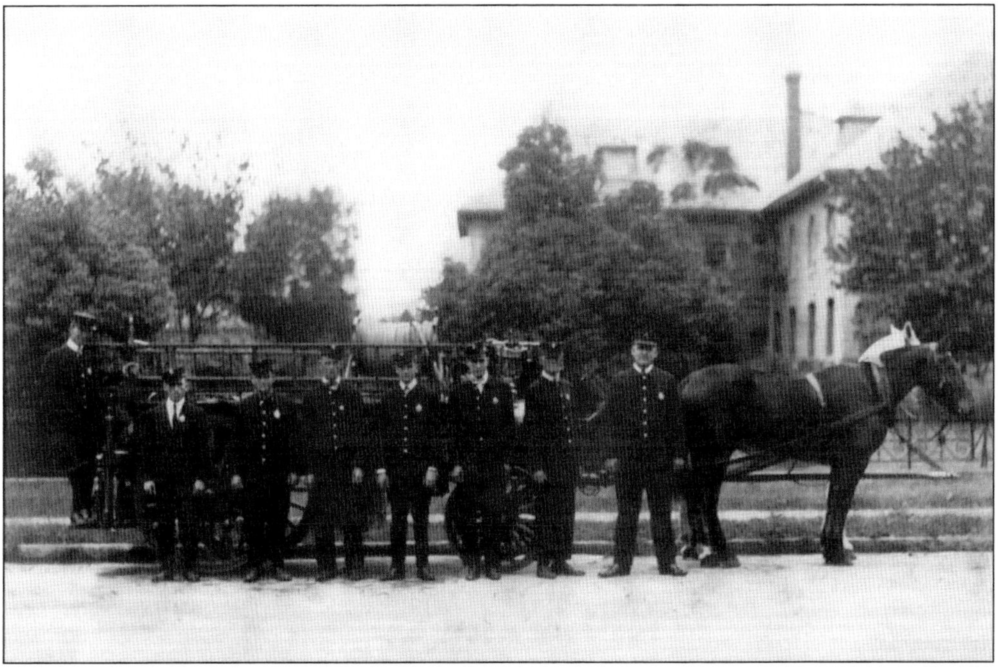

For its first 25 years, the HFD, as its predecessors had done since the 1830s, responded to fires with horse-drawn apparatus. The last horses in the department were retired to Hermann Park Zoo in 1924. Firemen continued to make the daily trip to the park to feed and groom the animals as long as they lived. The fireman at far left is J.H. Davidson, who later became chief.

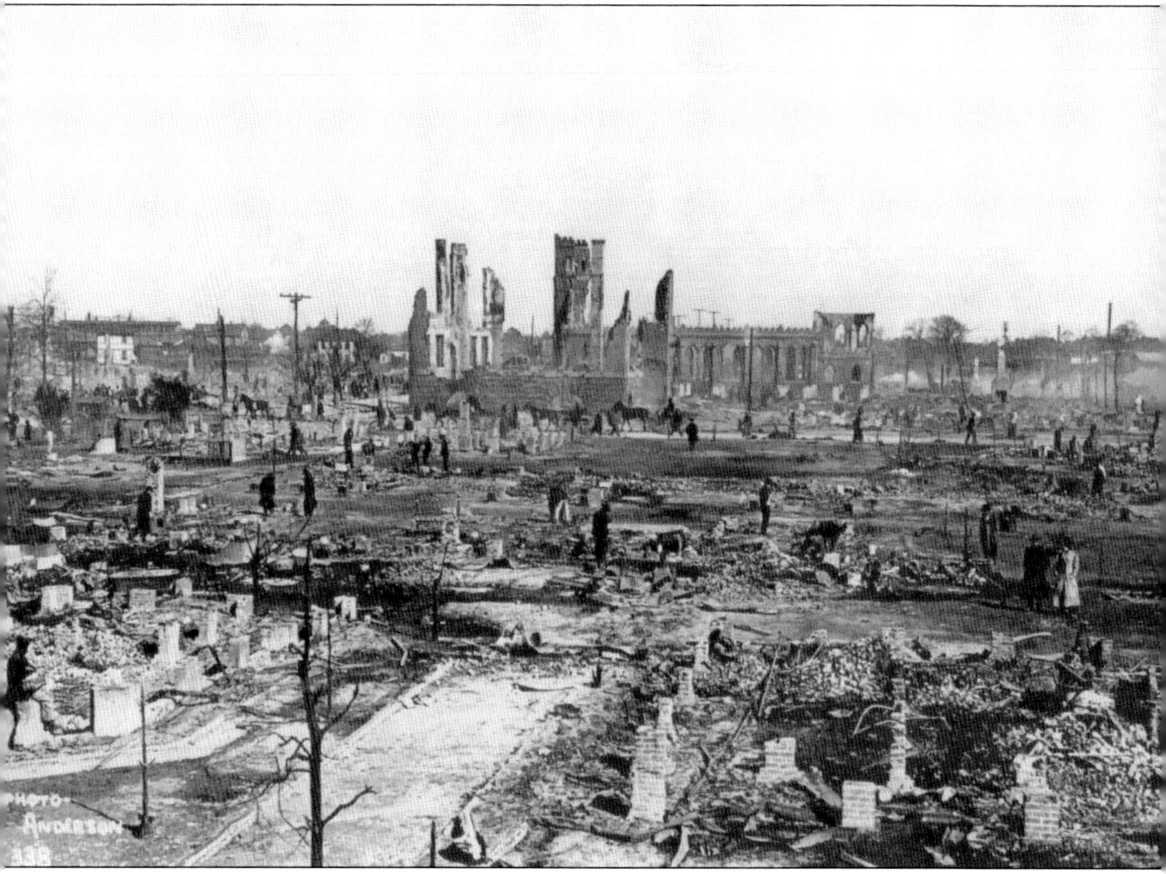

On February 21, 1912, aided by stiff northern winds blowing in at nearly 40 miles per hour, the largest fire in Houston's history began. It became known as the Great Fifth Ward Fire. The strong winds spread the blaze as embers set wood-shingled roofs on fire, a glow seen by train passengers nearly 35 miles away. Before the fire was out, it had incinerated a 40-block swath of Houston's Fifth Ward. It consumed St. Patrick's Catholic Church, a school, 13 industrial plants, 116 boxcars, nine oil tanks, eight stores, and 119 homes, mostly in the Fifth Ward. There were no deaths, but there was over $7 million in damage to residential and business property. The fire was started by transients in a vacant house attempting to keep warm, and traveled more than a mile and a half, crossing Buffalo Bayou and reaching the Houston Packing Company. Numerous large companies suffered, but the biggest hit came to the Cleveland Compress Company, which sustained a $2 million loss—over 32,000 bales of cotton smoldered for several days in their sheds and adjoining railroad yards. Mayor Baldwin Rice dispatched 12 city health inspectors the following morning to go to the damaged areas and assess resident's needs, as did a number of businesses and the United Charities organization. Public schools offered temporary shelter to residents displaced by the fire.

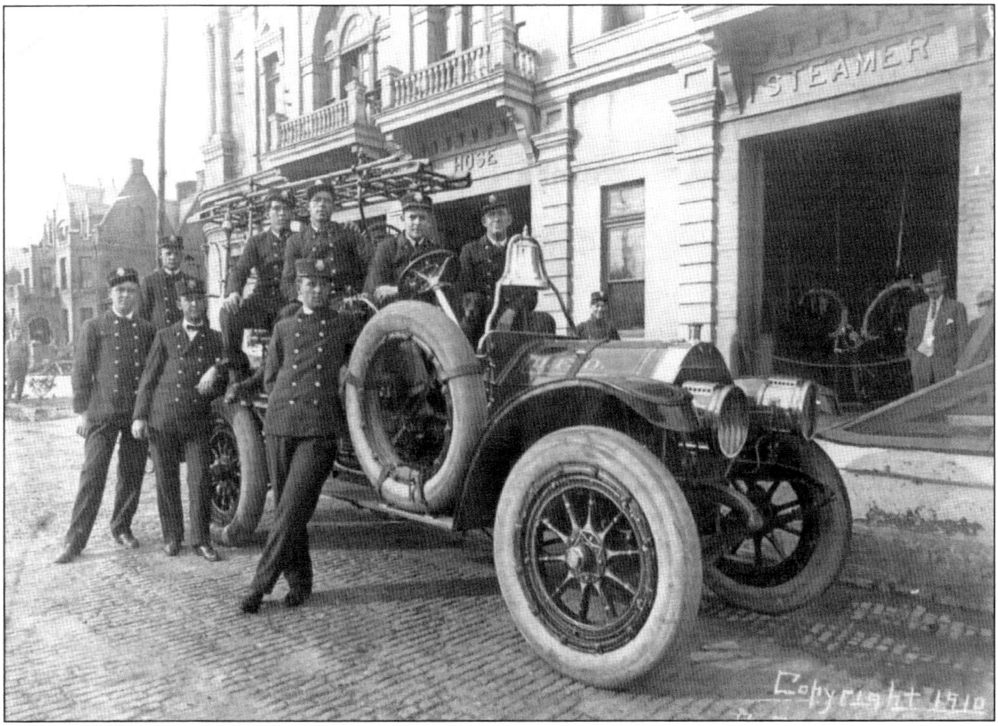

This chief's car was motorized in 1910, the first one for the department. It was the only motorized vehicle when Fred C. Seibert took office as fire chief in 1912. By the end of his first administration in 1919, most of the horse-drawn rigs in the department had been replaced by motorized vehicles.

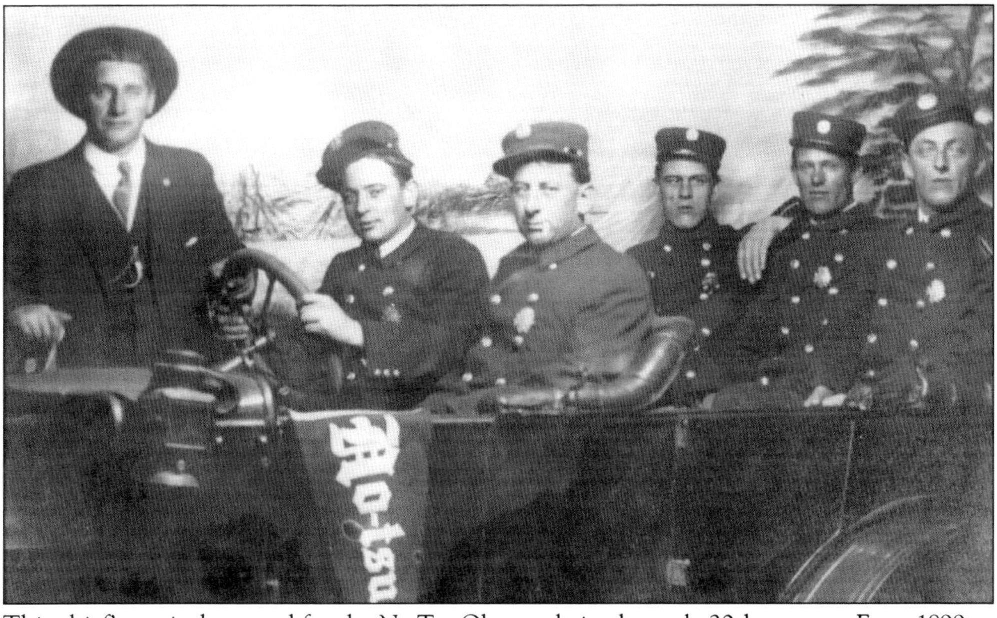

This chief's car is decorated for the No-Tsu-Oh parade in the early 20th century. From 1899 to 1915, Houston hosted an annual weeklong carnival, No-Tsu-Oh ("Houston" spelled backward), presided over by King Nottoc. The carnival was held in November to coincide with the football game between the University of Texas and the Agricultural and Mechanical College of Texas.

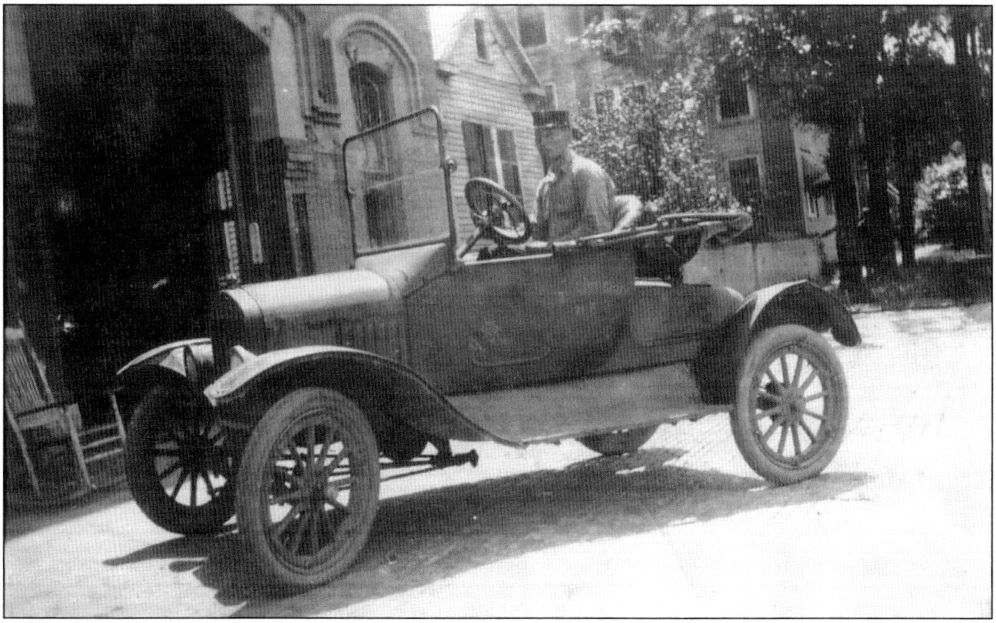

This second assistant chief's Model T Ford is parked in front of Station 6 in 1915. The station was constructed in 1903 at 1702 Washington Avenue and Ash Place. The single-bay, two-story brick building replaced the station built in 1892 by Mechanic No. 6 volunteer fire company at 1106 Washington Avenue. It was replaced in 1931 by a building at 901 Henderson Street, which was itself replaced in 1987 by a station at 3402 Washington Avenue. The first motorized apparatus, a chemical engine, was built by the department's shop and completed in 1915. By 1921, all of Houston's fire trucks were motorized. Most of the first-line apparatus were chain-driven motor vehicles. Many reserve apparatus, such as the steamers and the department's only water tower, were formerly horse-drawn rigs harnessed to motorized tractors.

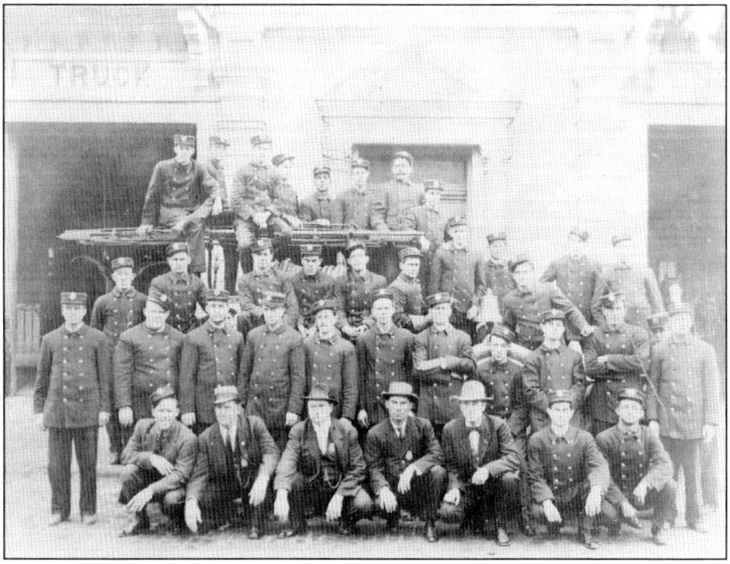

This is a 1911 photograph of the men assigned to the crew at Central Station, located at 519 San Jacinto Street. Fire Chief Fred Seibert is second from left in the first row. The original central station, built in 1894 for the volunteer Hook & Ladder Co. No. 1, was destroyed by fire in early 1904. Operations would quickly outgrow this building, which was replaced in 1924.

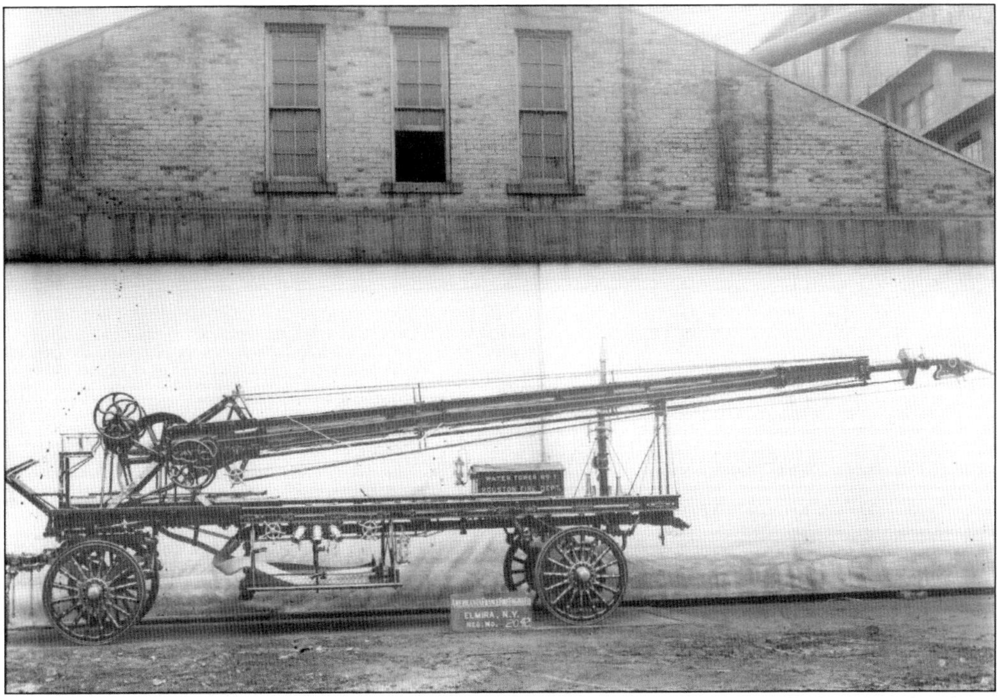

Houston's only water-tower apparatus was a 1912 American Automatic, pulled to fires by three horses. By 1923, the vehicle had been redesigned as a single-chassis truck with an American LaFrance tractor. The tower was housed at Central Station and responded to large downtown fires until 1966, when the department took delivery of two snorkels.

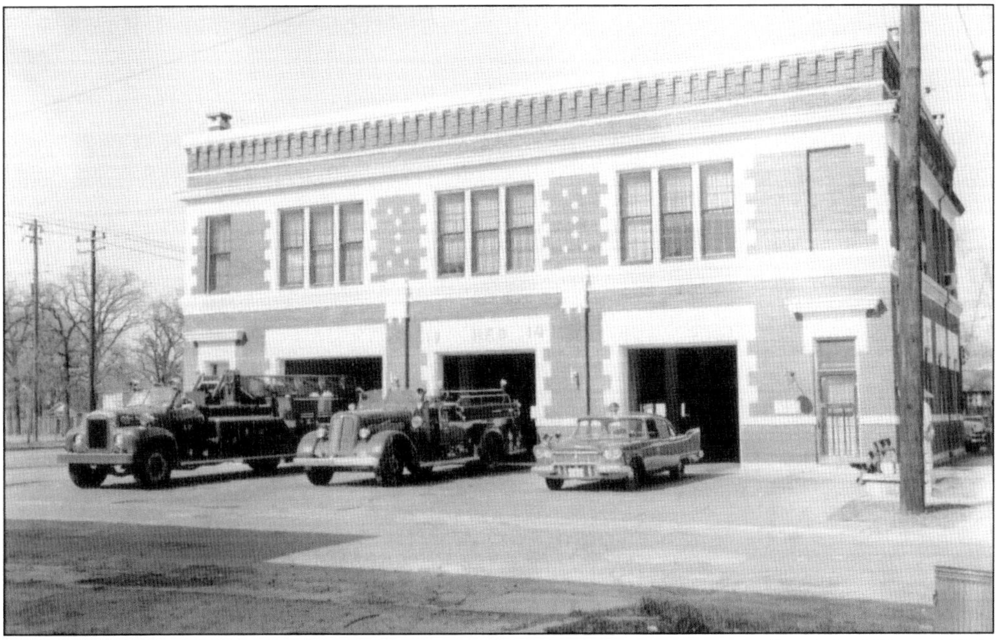

Station 14 was once the oldest active station in the department. It was built at Twelfth and Yale Streets as the municipal building for the city of Houston Heights in 1914 before the area was annexed by Houston in 1918. The old jail in the building served as the firefighters' gear room.

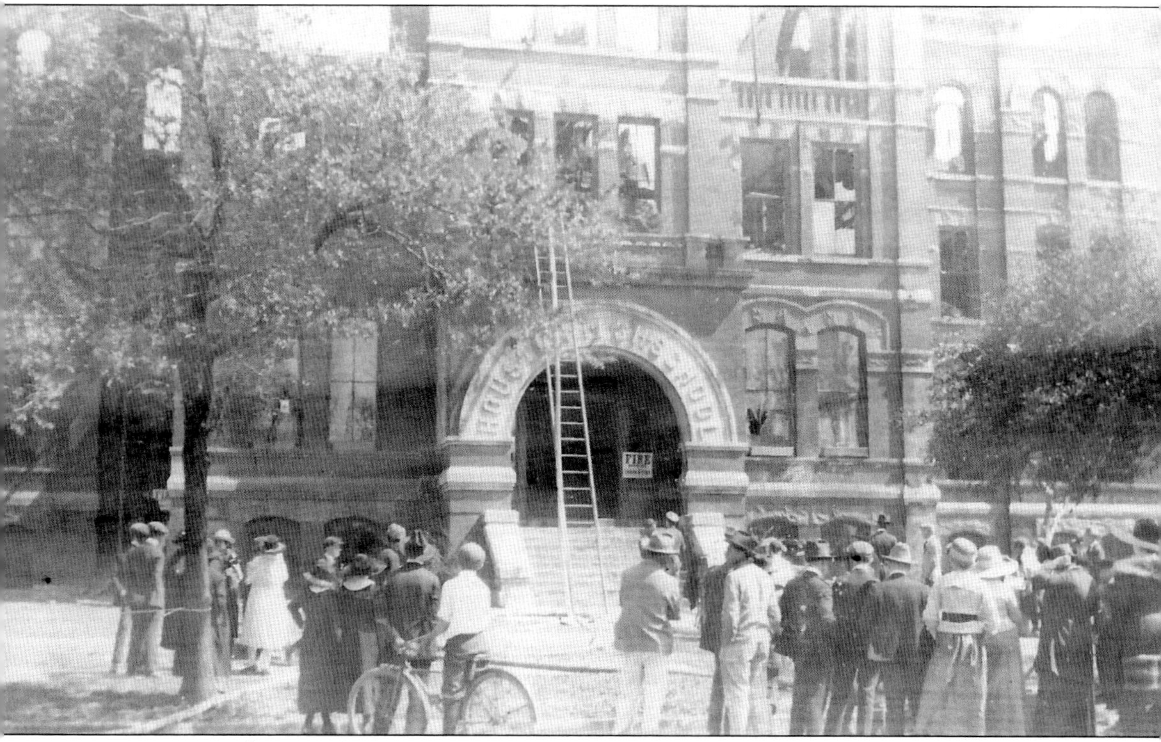

Spectators gather outside Central High School at Rusk and Austin Streets during a fire on the morning of March 18, 1919. Central High, which broke ground in 1894, had been described as one of the finest high schools in this part of the country. The school was gutted by this fire, with damage estimated at $175,000. Sam Houston High School replaced Central High at that site in 1921.

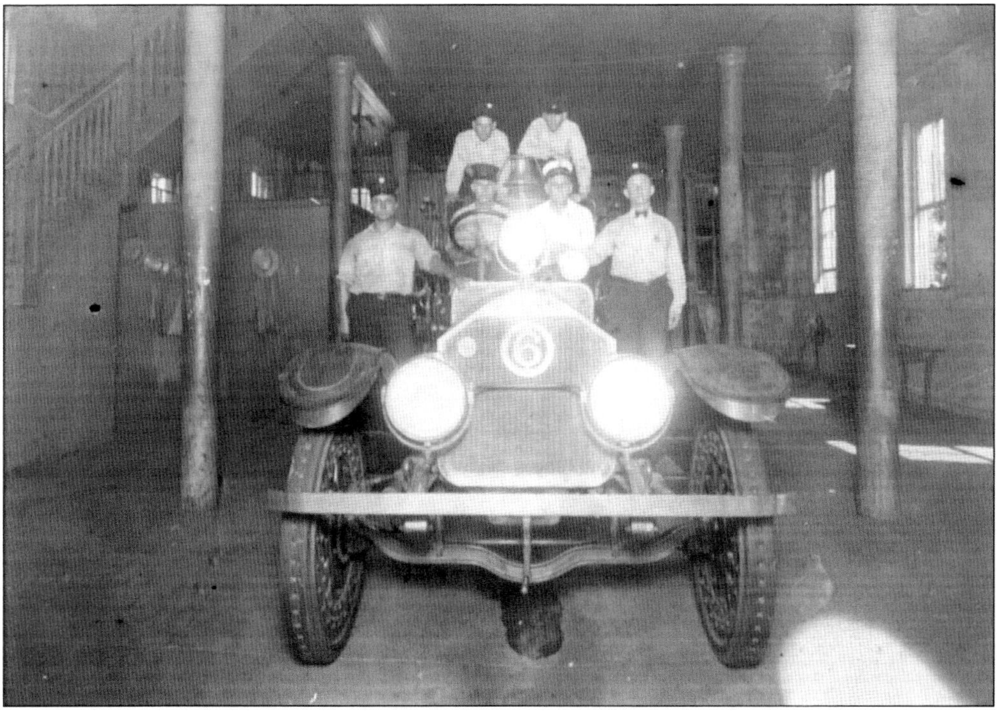

This 1919 photograph shows the crew of Station 6 at 1702 Washington Avenue. The photograph includes the station's American LaFrance engine and offers a great view of the apparatus bay. Many of the photographs of firefighters and apparatus were taken outside. The man in the white hat is Capt. John Little, who lost his life in the line of duty in 1929.

A rare snow, a full two-tenths of an inch, provided ample material for the crew of Central Station to create a snow-woman with the aid of neighborhood children on January 11, 1918. Snow in Houston is unusual. Light snow has fallen approximately 35 times since 1895, when 20 inches fell on February 14–15. The rectory of Christ Church Cathedral is in the background.

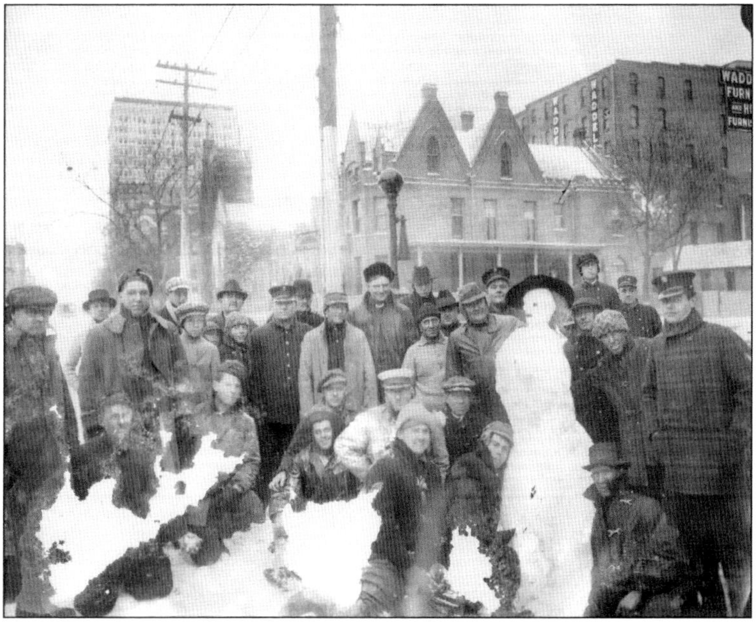

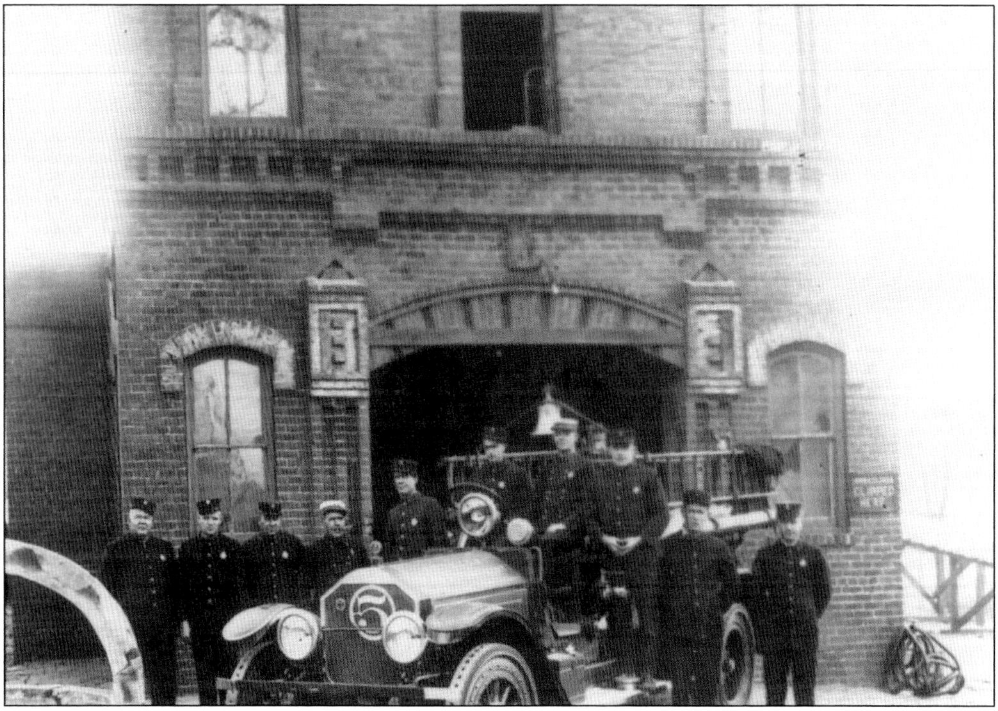

Station 5 opened at Liberty Road and Nance Street in 1874 to house Brooks No. 5. The company folded in 1883, but reformed under the paid fire department at the turn of the 20th century. A new Station 5 was described as including hot and cold baths, a unique feature for firefighters at that time. This 1920s crew poses outside the station, located at 910 Hardy Street at Nance Street.

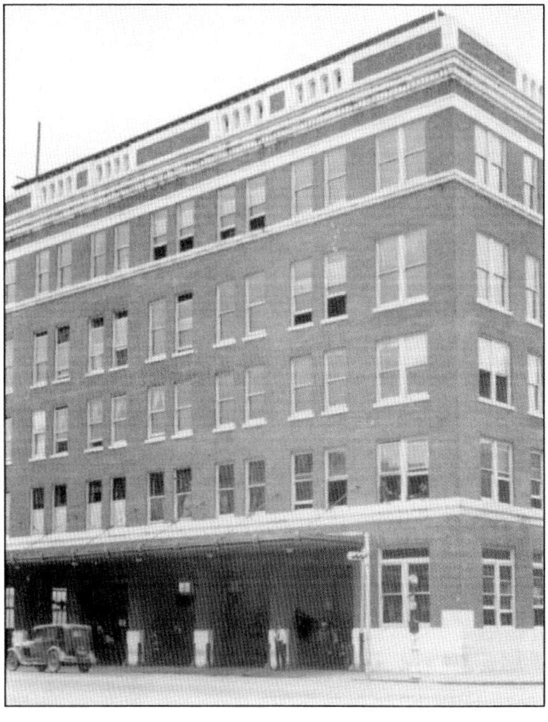

A new five-story Central Station was constructed in 1924 at the corner of Preston and Caroline Streets. Administrative offices and the dormitory were on the second floor. Police headquarters occupied the upper three floors, and the fire alarm was located in the penthouse. The building is used today as a courthouse annex for Harris County. The apparatus bay doors have been cemented over, but the structure remains virtually unchanged otherwise.

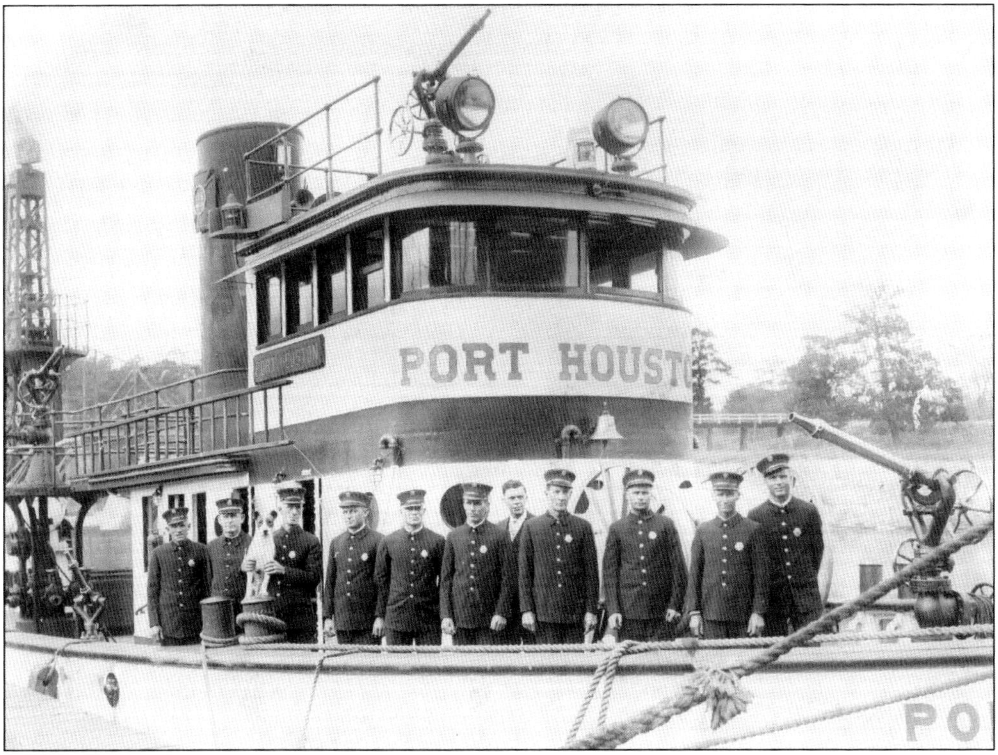

Port Houston was put into service in 1926. The ship's cost, $300,000, was provided by funding from the city, Harris County, and the Navigation District. There were 18 men assigned to the boat, which was docked at City Wharf No. 5 inside Station 21. The boat could throw a dozen streams simultaneously and had a pumping capacity of 7,000 gallons of water per minute.

From 1924 to 1940, the fire alarm office was in the Central Police and Fire headquarters at Preston and Caroline Streets. The department had 238 men and, in 1924, answered 2,023 alarms. The number of alarms answered grew as the city expanded. As Houston annexed new areas and existing communities, strategic planning was utilized to ensure that public safety was handled properly.

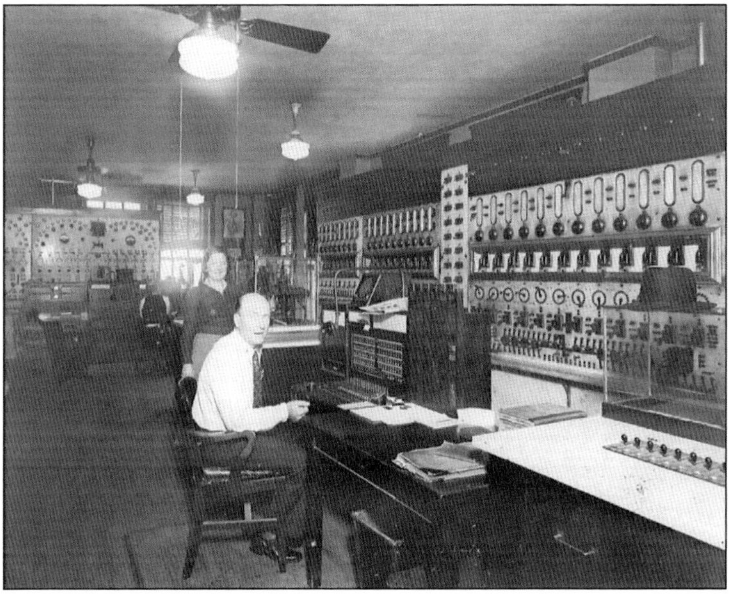

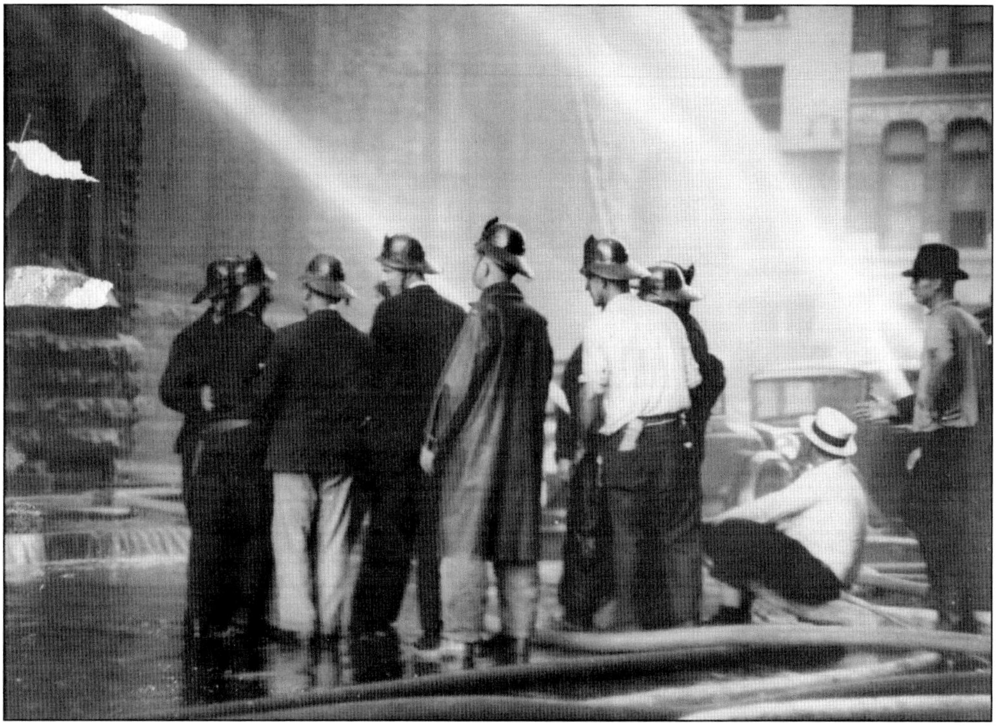

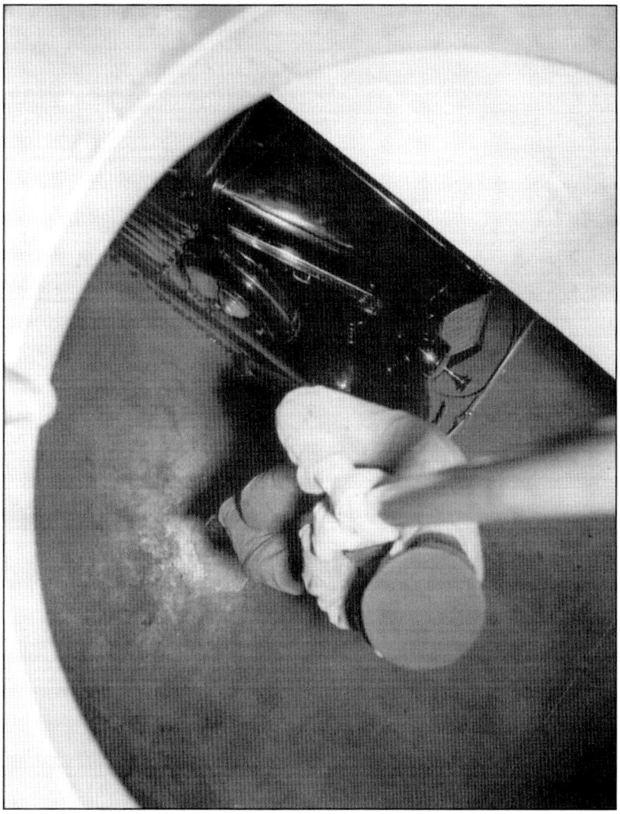

Firemen's protective gear in the 1930s consisted of a helmet made of a stiff, cardboard-like material, shirtsleeves, a blue serge coat or a heavy raincoat, and boots. In the late 1970s and early 1980s, the National Fire Protection Association adopted standards for protective clothing, gloves, and SCBA. As the fire service grew and improved technologically, the gear evolved.

It was not until the 1930s that the fire department began issuing helmets to the firefighters as part of their uniforms. Previously, Houston firefighters were required to provide their own. The first ones were made of a composite material that was essentially cardboard heavily coated with shellac and painted black.

Two

The Motorized Magnolia City

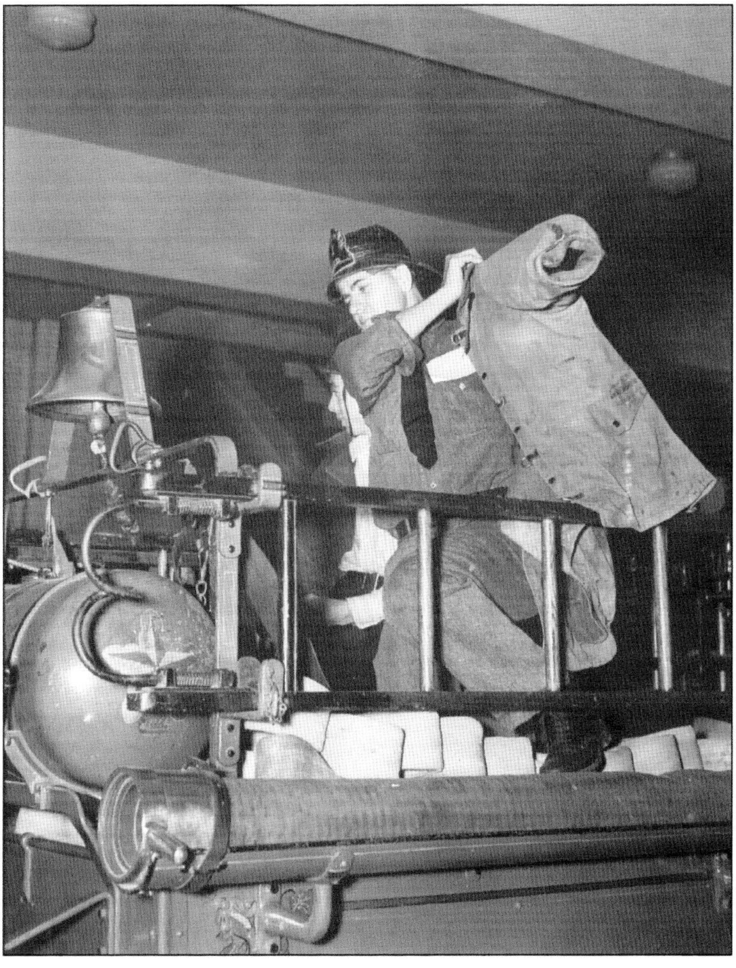

By the 1930s, the Houston Fire Department had 382 firefighters assigned to 26 fire stations throughout the city. Apparatus similar to the one pictured here consisted of twenty-six pumpers, eight ladder trucks (including two aerial ladders), a water tower, and three chemical wagons.

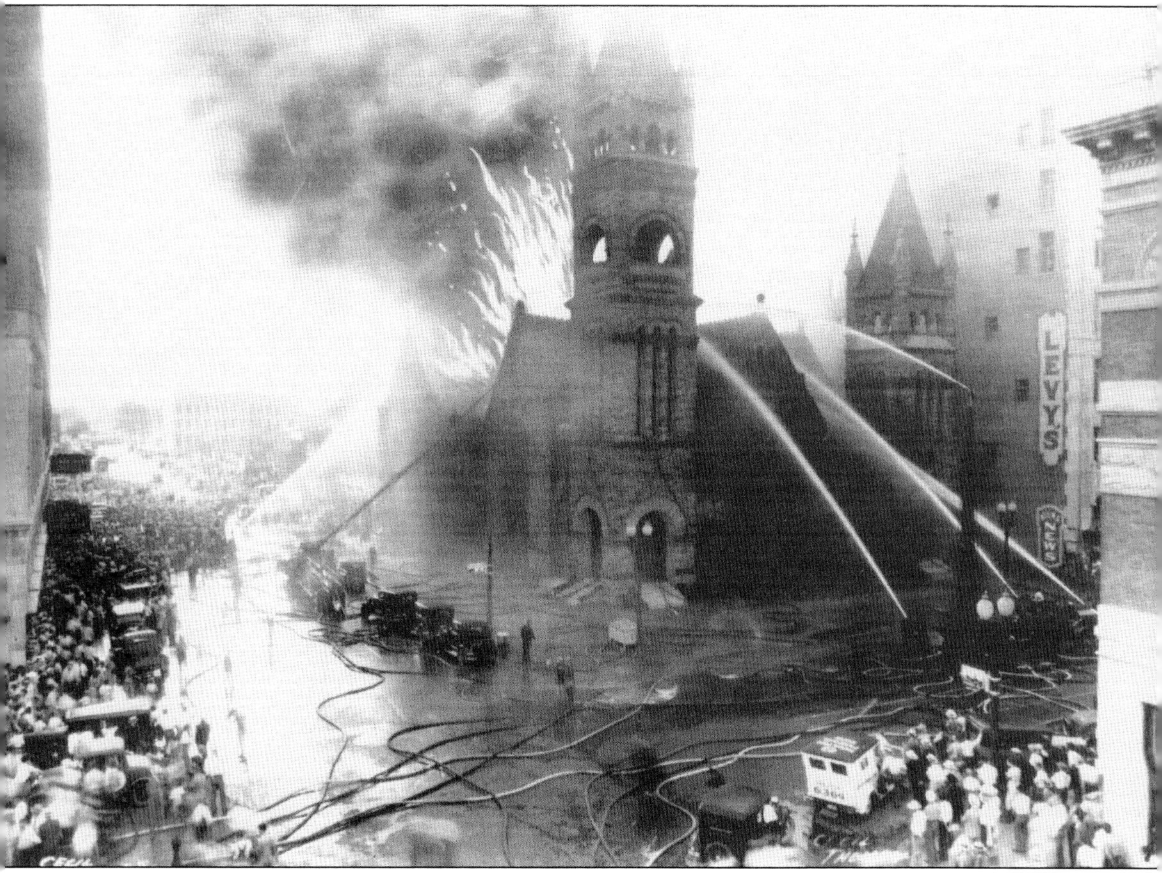

The First Presbyterian Church, located downtown at Main and McKinney Streets, caught fire on September 10, 1932. The flames shown in this photograph were embellished by the *Houston Post*. The fire caused over $100,000 in damage and temporarily displaced the congregation. Services were held at the neighboring Metropolitan Movie Theater until the church was rebuilt.

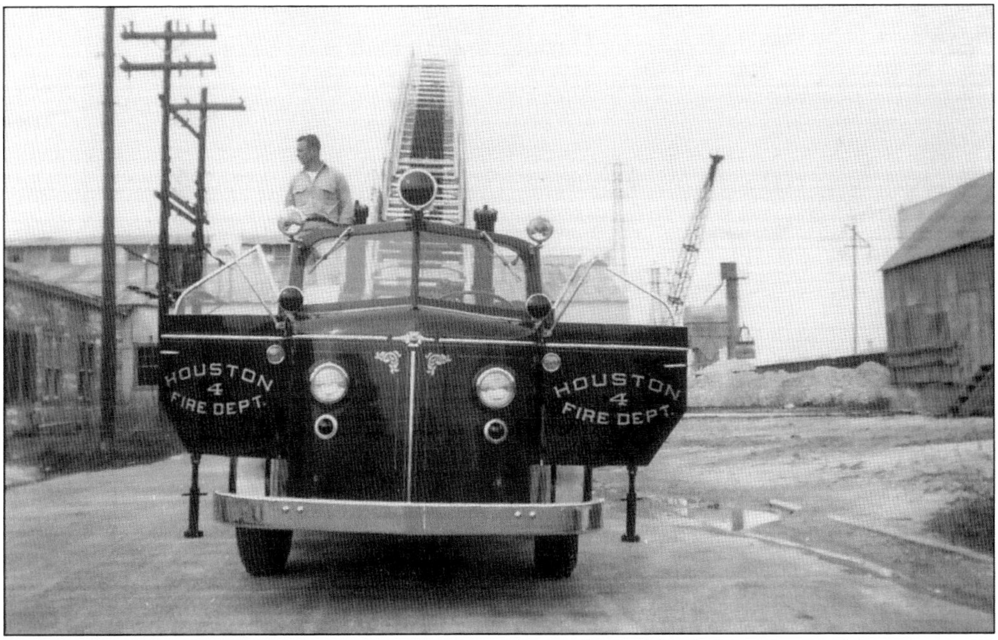

Ladder trucks, like this one from Station 4, have long allowed firefighters to quickly gain access to the upper reaches of tall buildings. Metal aerial ladders were introduced in the 1930s. Aerial ladder trucks make for quicker rescues and the application of large amounts of water through the large hose and nozzle devices attached to the tip of the apparatus.

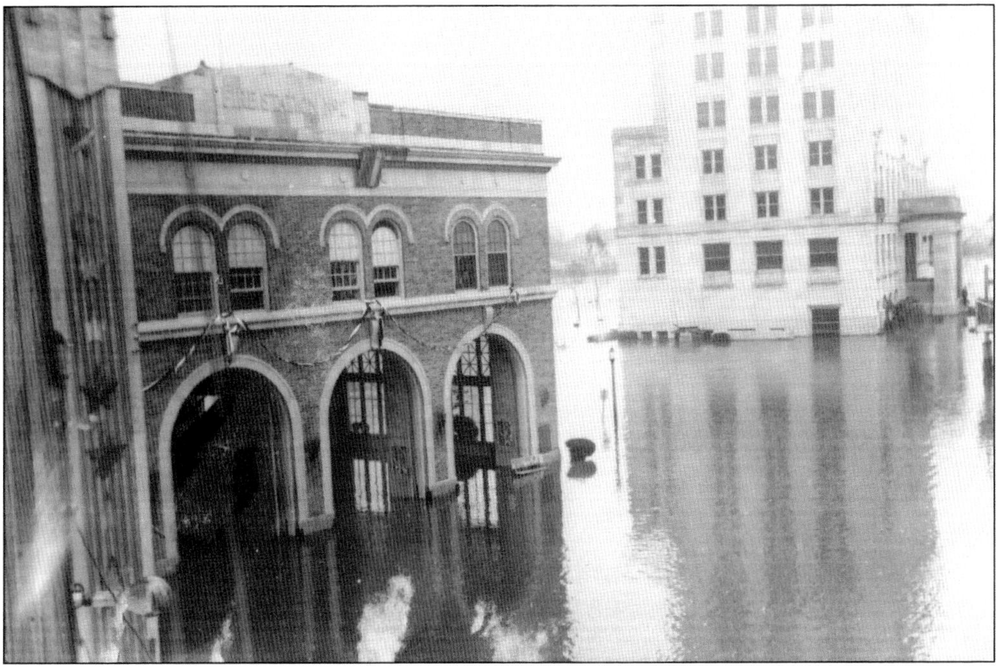

A December 1935 flood raised the waters of Buffalo Bayou eight feet and affected downtown areas. A garage of the Yellow Cab Company caught fire, and the only way to get water to fight the blaze was to draft from the flooded streets. The floodwaters surrounded Station 2 (pictured). Canoes became a secondary form of transportation for most of downtown.

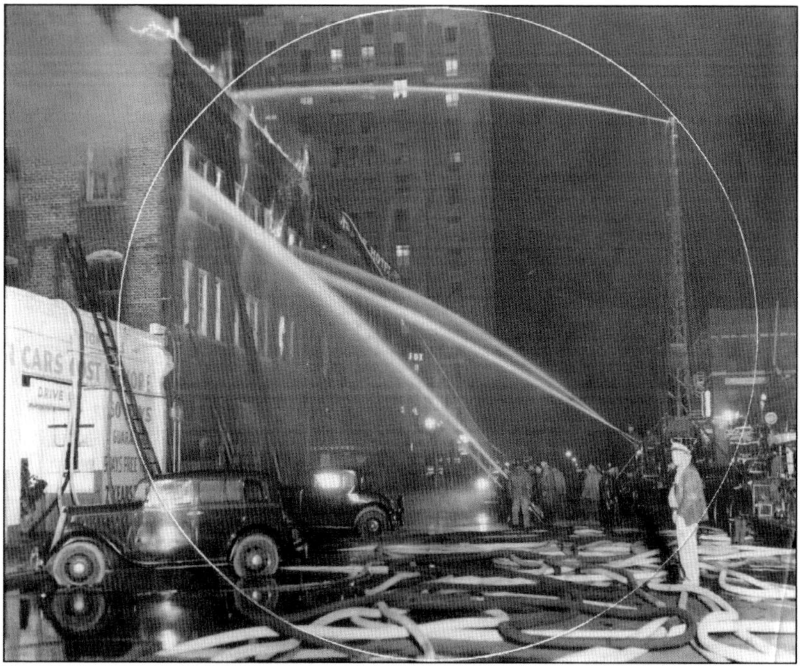

HFD fought three large fires in December 1935. The blaze at the Thomas Goggan & Brothers Music Store was the largest. The four-alarm fire started the evening of December 26. The building's first two stories housed the music store, and the top floor was the Fox Hotel. All of the guests were safely evacuated.

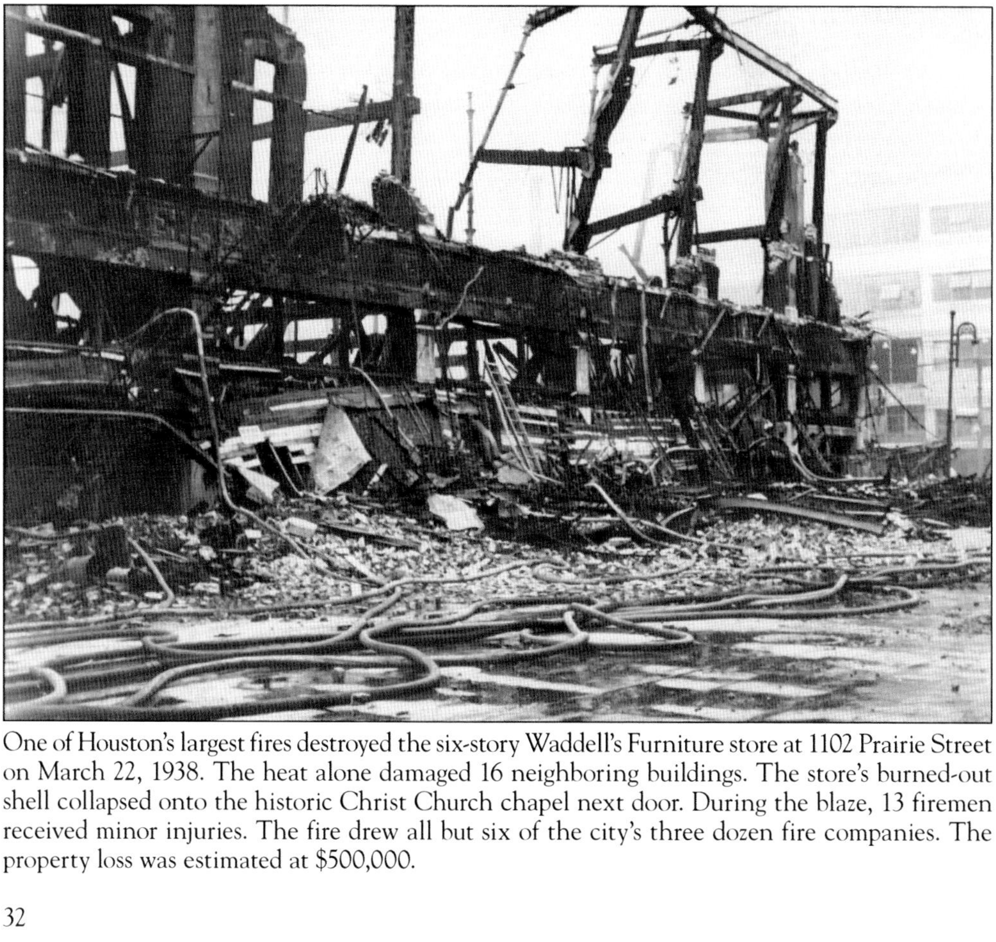

One of Houston's largest fires destroyed the six-story Waddell's Furniture store at 1102 Prairie Street on March 22, 1938. The heat alone damaged 16 neighboring buildings. The store's burned-out shell collapsed onto the historic Christ Church chapel next door. During the blaze, 13 firemen received minor injuries. The fire drew all but six of the city's three dozen fire companies. The property loss was estimated at $500,000.

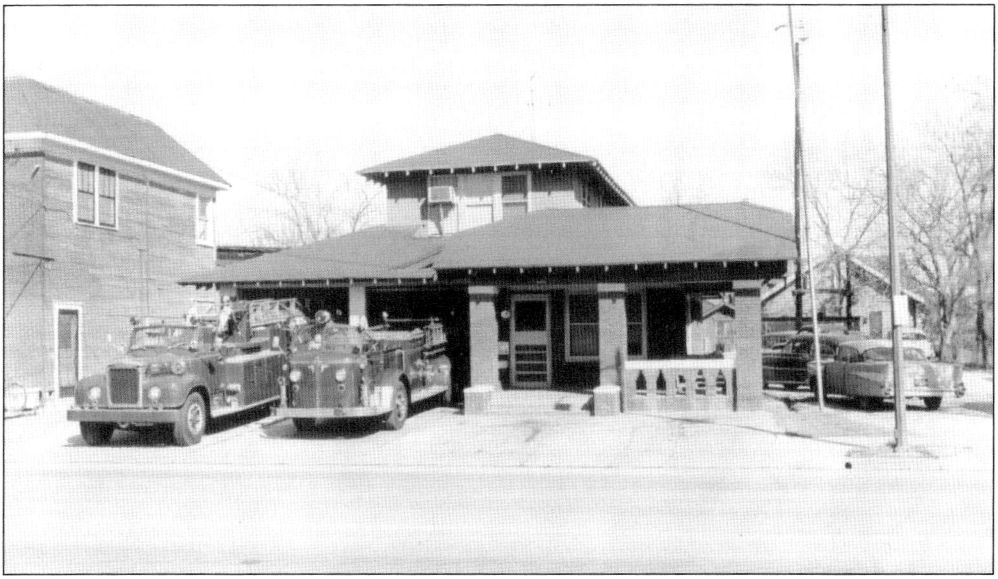

Station 22, at 7825 Harrisburg Boulevard, was brought into the department after a 1926 land annexation. Residential fire station design changed dramatically between 1905 and 1940; facilities began resembling the homes they protected. Like single-family homes, "bungalow" stations were set back from the street, had landscaped lawns and gardens, housed apparatus in an attached garage, and featured a front door covered by a porch.

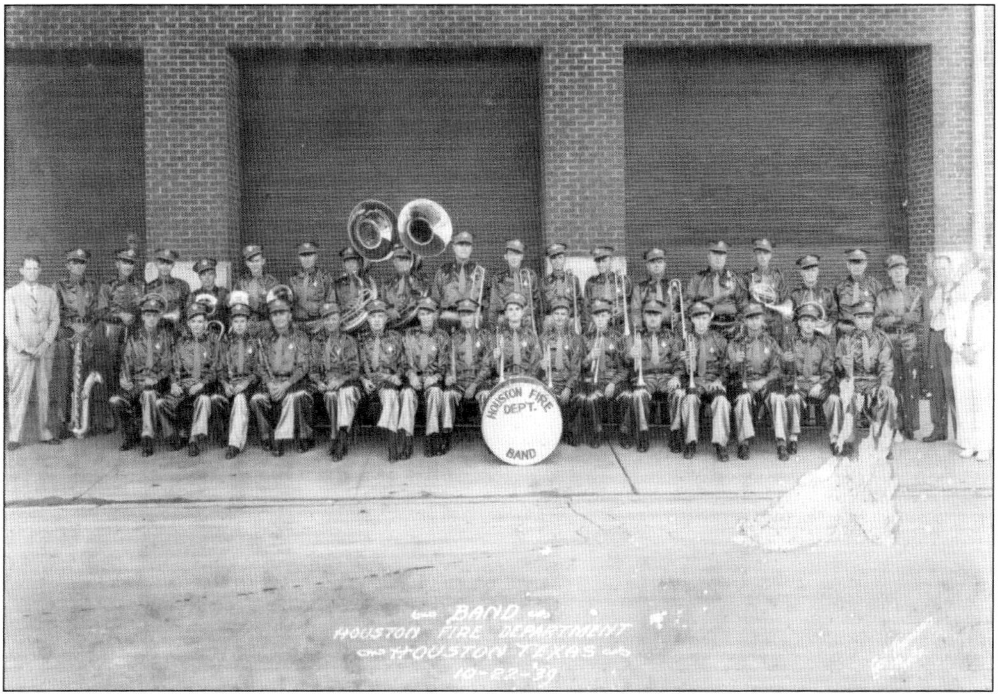

The two men in suits at either side of the HFD band (pictured in October 1939) are Fire Commissioner Frank Mann (left) and Fire Chief Homer Lyles (right). Commissioner Mann was instrumental in the organization of this band. There was also a Fire & Police Band under the direction of Tom Fennell. Many men played in both bands.

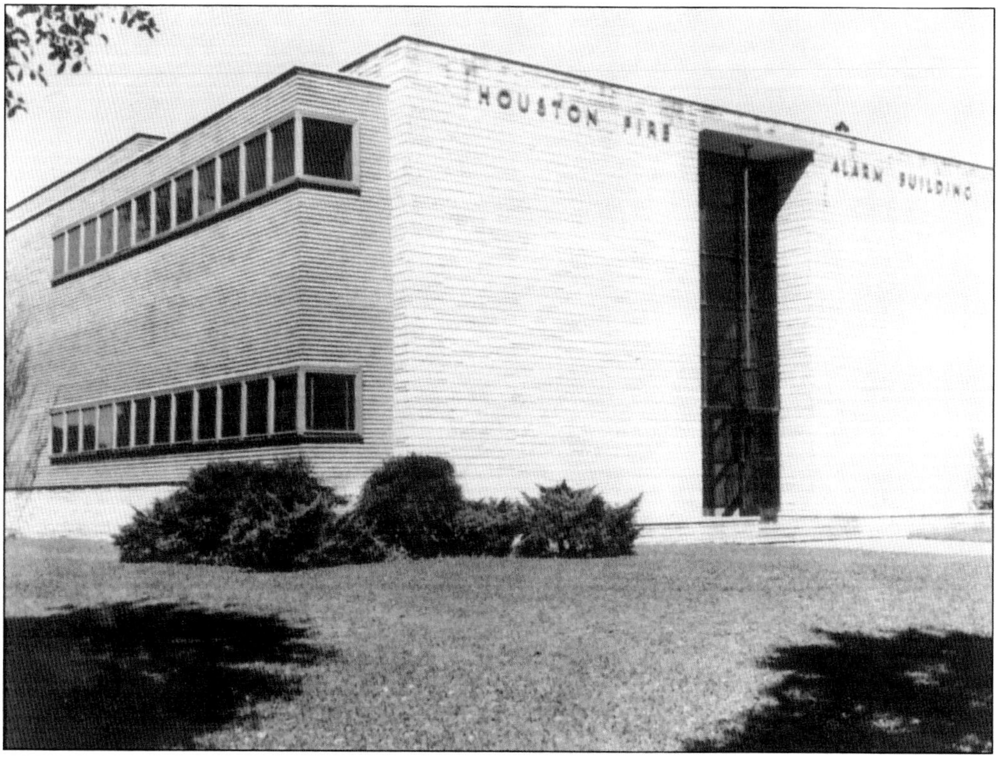

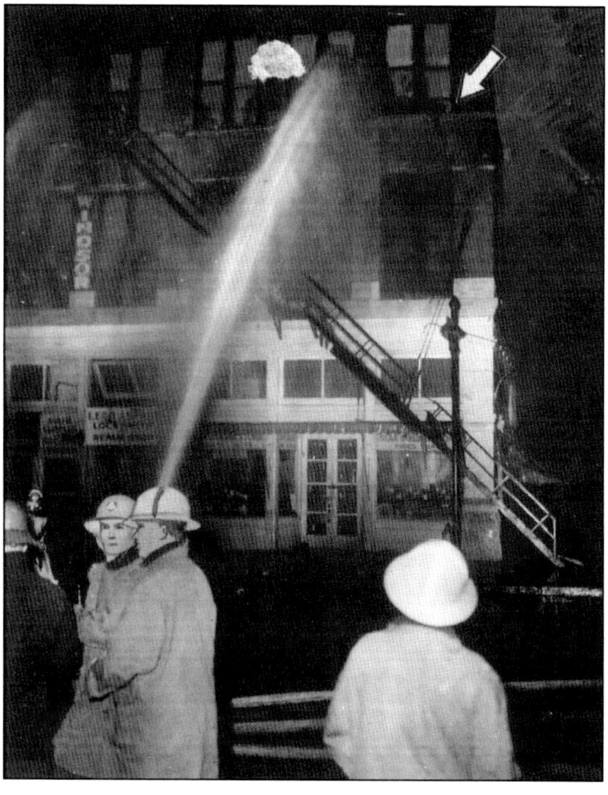

In the spring of 1940, Houston's fire-alarm office received a home of its own, at Bagby and Lamar Streets. All of the fire-alarm equipment was relocated, and the number of Gamewell fireboxes increased to 820 units. Dispatchers in the new office used a vocal alarm system to alert the city's fire stations. The system was connected to the central facility via a station loudspeaker with the flick of a switch.

The worst loss of life in a single fire in Houston's history occurred on Tuesday, September 7, 1943. In the blaze that destroyed the Gulf Hotel at 615 Preston Street, 55 people were killed and dozens were injured. The fire reportedly started in a burning mattress, thought to have been ignited by careless smoking. The arrow in the photograph indicates a person stuck in a window.

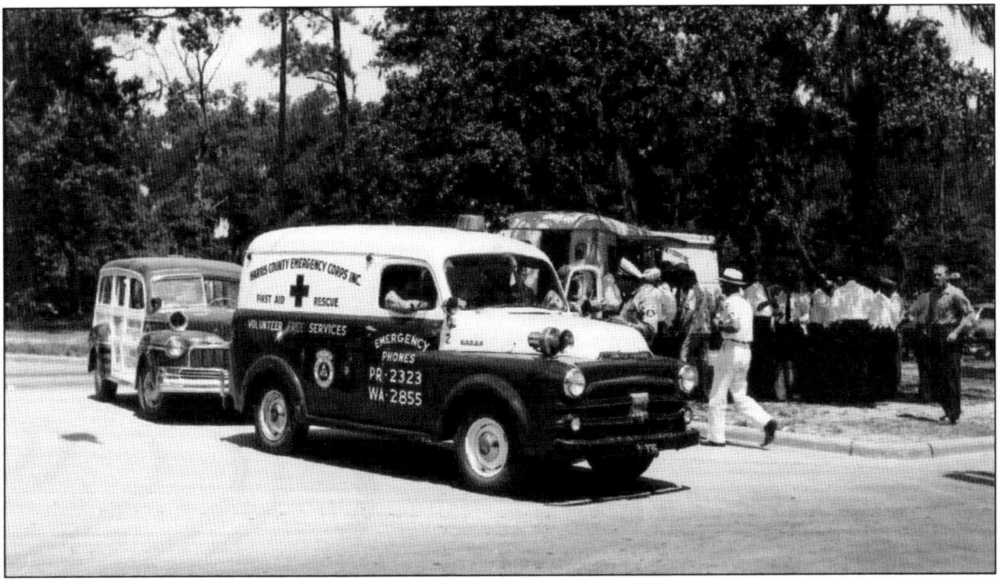

Ted Felds founded Harris County Emergency Corps Inc. in 1939 and taught first aid classes to hundreds of firefighters and many community groups and city departments. In over four decades, Felds's group taught over 10,450 people—more than any other volunteer instructor for the Red Cross in the United States.

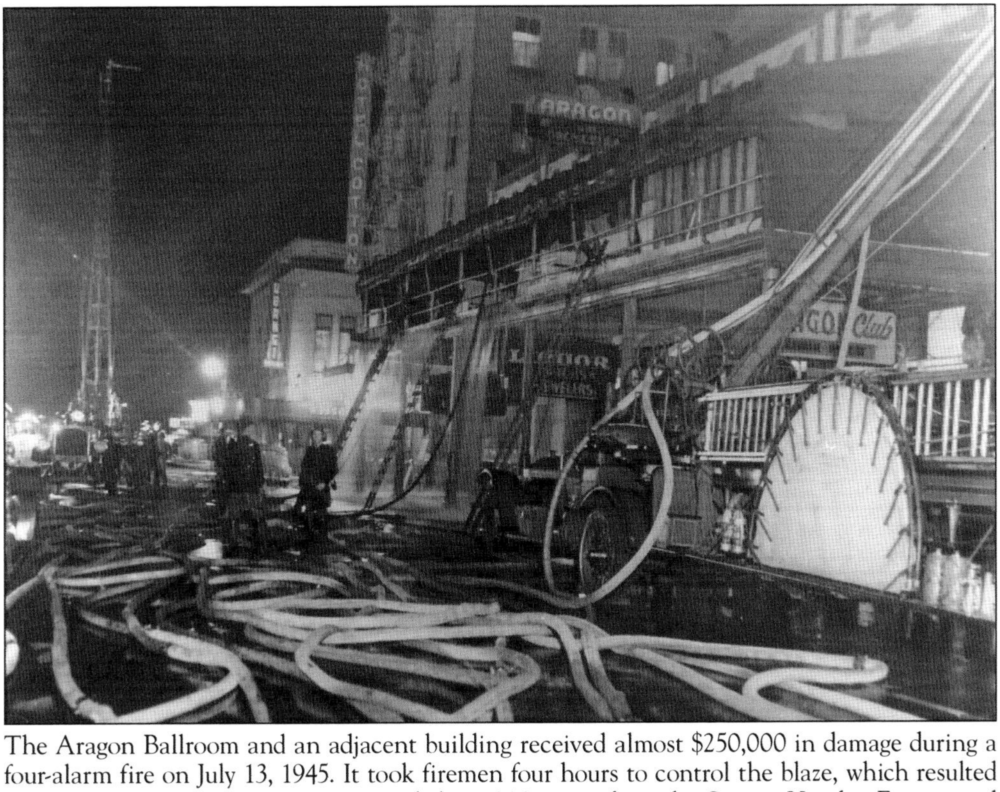

The Aragon Ballroom and an adjacent building received almost $250,000 in damage during a four-alarm fire on July 13, 1945. It took firemen four hours to control the blaze, which resulted in dozens of injuries. Firemen evacuated about 200 guests from the Cotton Hotel at Fannin and Rusk Streets during this fire.

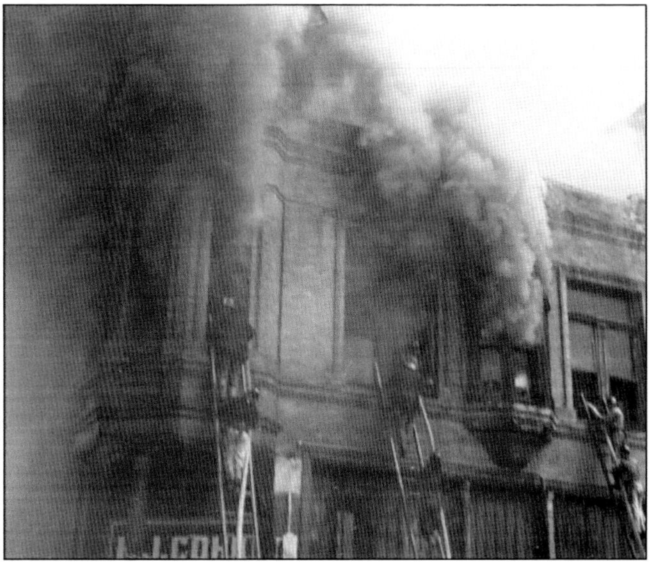

On the afternoon of June 25, 1946, the fire department was called to the H.J. Cohn Furniture Warehouse, located at 201 San Jacinto Street. The building was engulfed in flames when firefighters arrived. The contents of the store were completely destroyed, with an estimated damage of $40,000. During the blaze, 18 firefighters were injured. A young man, previously questioned for three other fires, was eventually arrested.

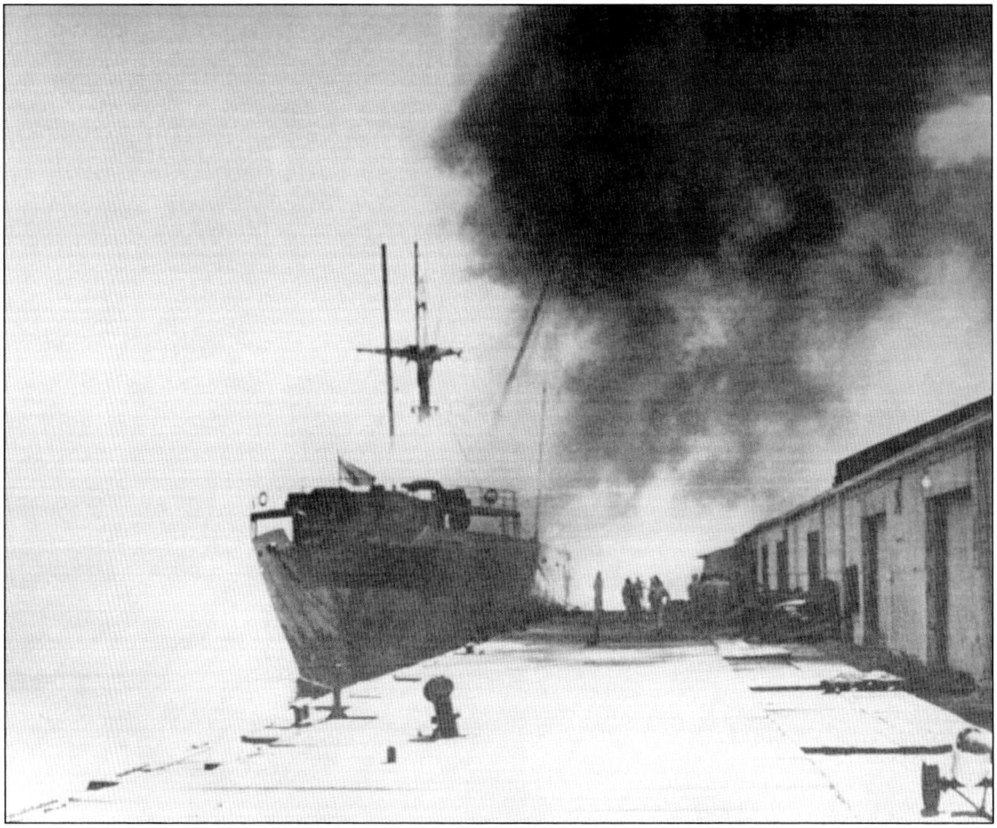

On Wednesday morning, April 16, 1947, a fire started in the hold of the French steamer *Grandcamp*, docked at Texas City. Stored in the hold were thousands of bags of fertilizer-grade ammonium nitrate. The Texas City Volunteer Fire Department responded and, with the help of longshoremen and several private-brigade firefighters, began to battle the fire. Spectators by the hundreds turned out to watch the action.

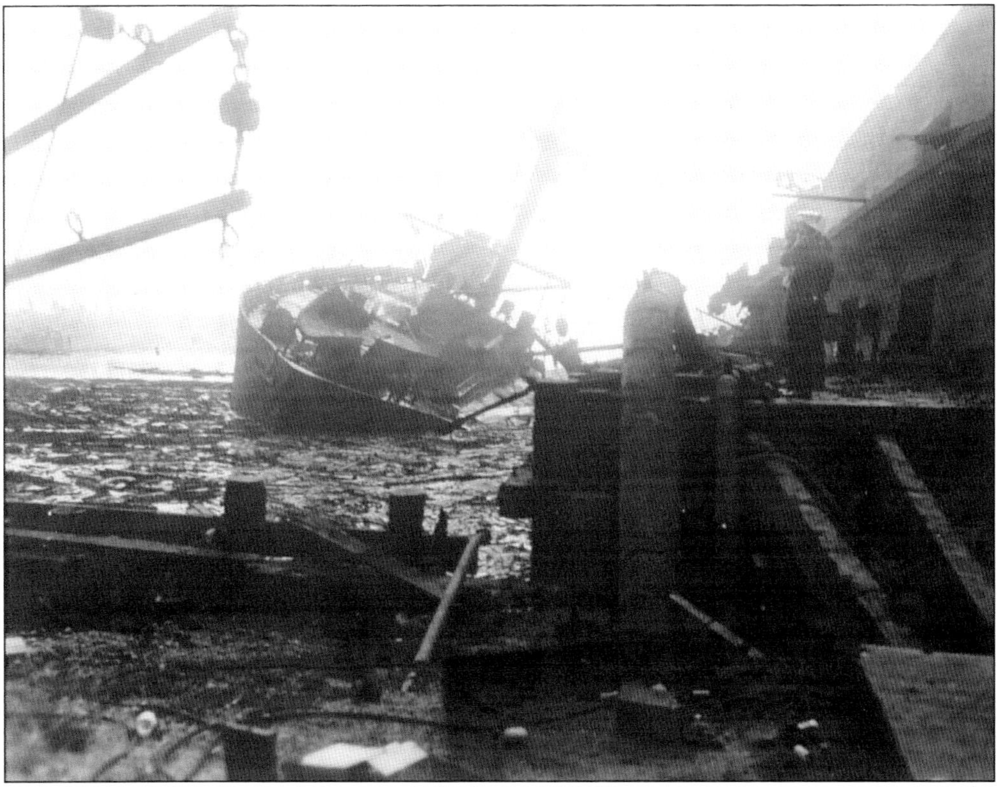

Shortly after 9:00 a.m., firefighters sealed off the ship's hatches in an attempt to smother the blaze. Just 12 minutes later, the *Grandcamp* exploded. The blast was heard and felt in Houston, 40 miles to the north. Flying debris started smaller explosions and fires throughout Texas City, and a 15-foot tidal wave swept the dock area.

Within a few hours, Houston Fire Department Engines 3, 17, and 27 arrived to assist the hundreds of rescue workers and volunteers already gathered and working to fight the *Grandcamp* fire. There was danger of another explosion; the entire dock area was in flames, as was the SS *High Flyer*, also filled with ammonium nitrate. Luckily, the dock area had been mostly abandoned; the following morning, the *High Flyer* exploded.

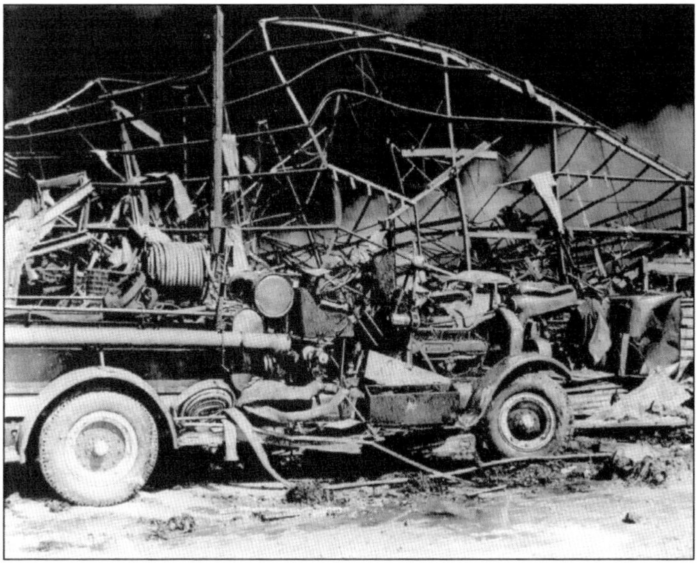

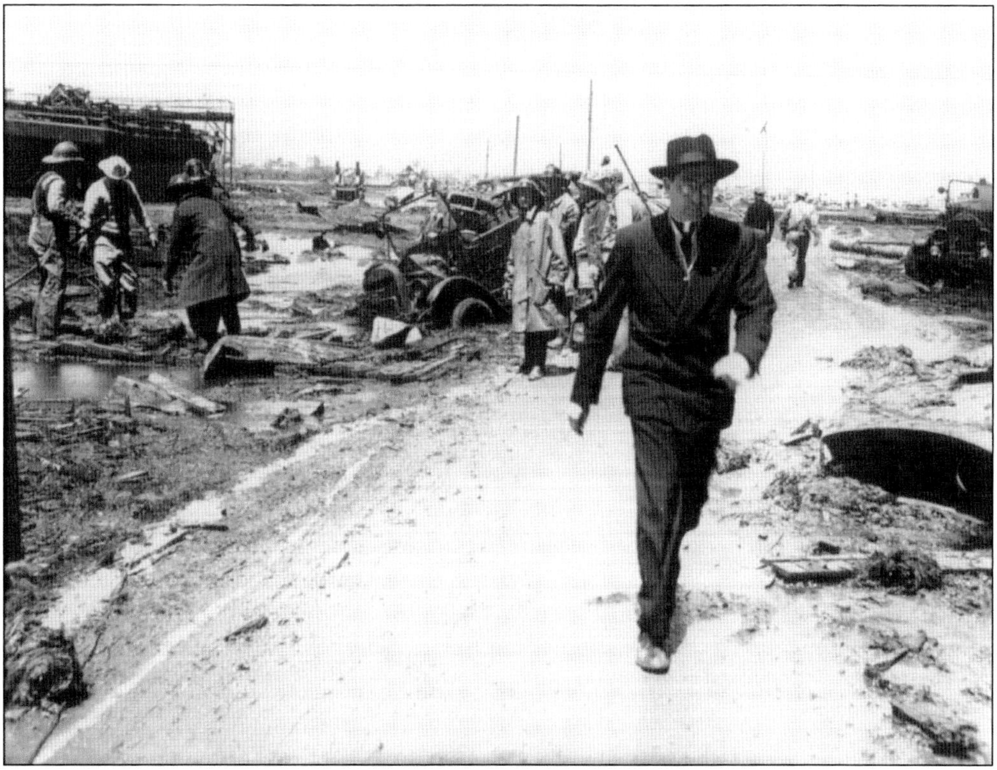

The statistics of the Texas City explosions were grim: 576 people killed and over 4,000 injured. All personnel of the Monsanto Chemical Company were killed, and all of the company's property was destroyed. Property damage from the *Grandcamp* explosion alone was estimated at $67 million. All 23 members of the Texas City Volunteer Fire Department and two members of the Texas City Heights VFD died.

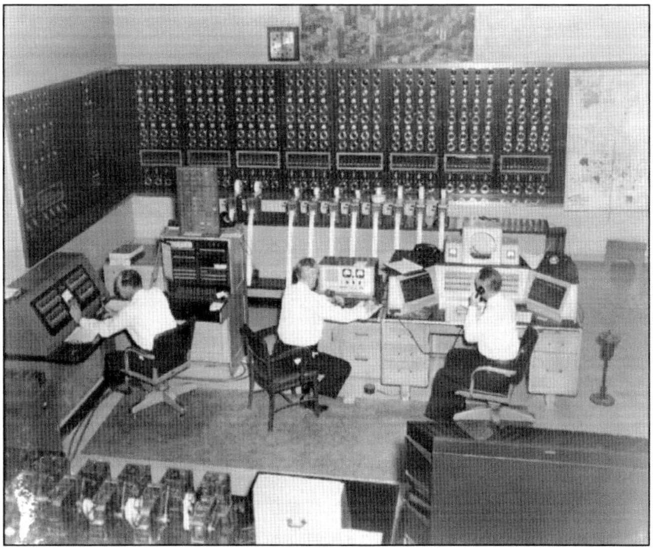

After World War II, the HFD's communications system improved dramatically with the addition of radios, first in district chiefs' cars, then in every first-line apparatus. In 1950, the fire-alarm division began recording all telephone-reported fires, allowing conversations to be replayed to verify an address, if needed. The following year, HFD received its own radio transmitter, allowing the department to separate its communications system from that of the police department.

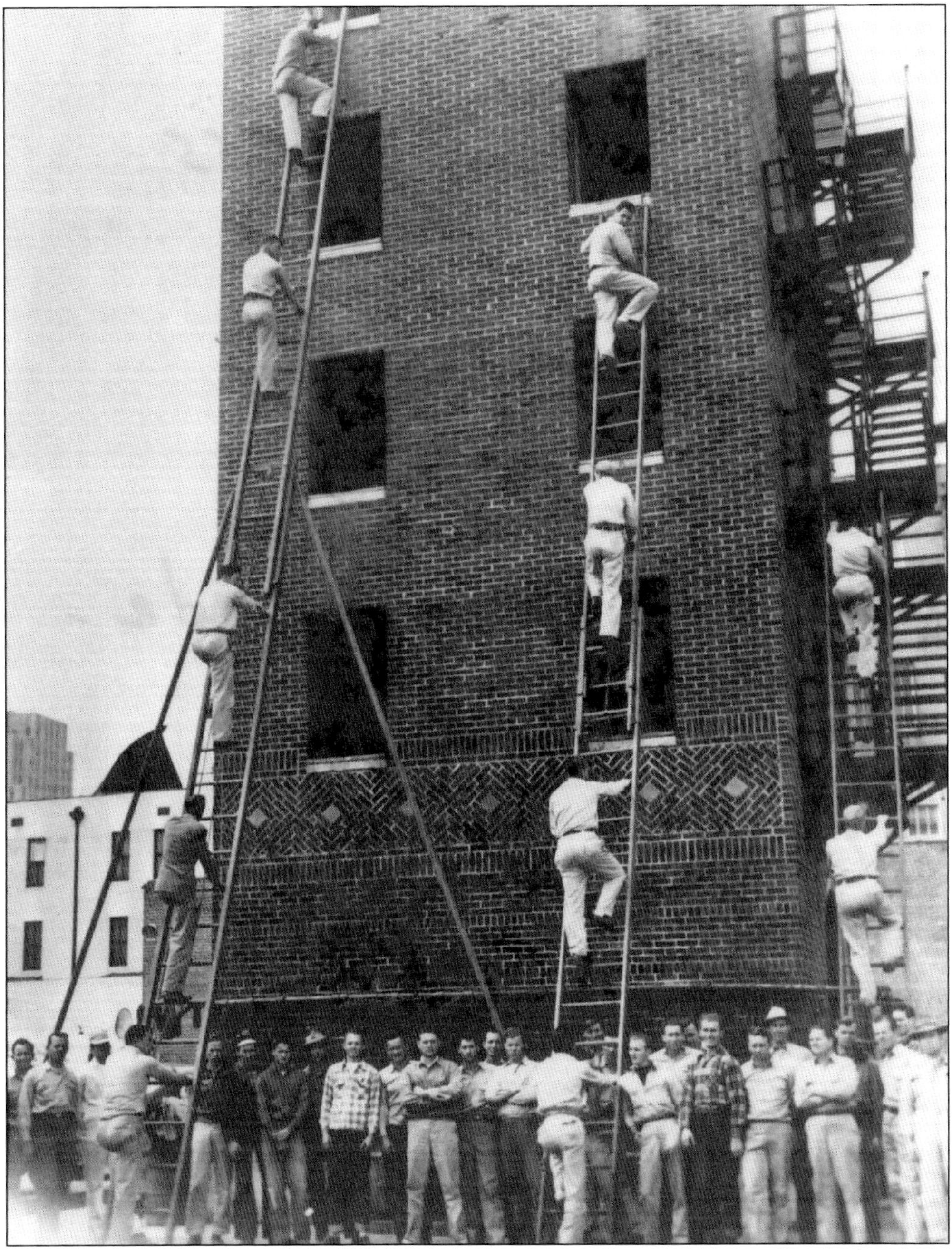

In 1928, a fire-department training facility was built beside the Central Fire Station at the corner of Preston and Austin Streets on a recommendation of the National Board of Fire Underwriters. The training facility had a five-story drill tower (shown in this 1950 photograph) with a smoke basement. A small building beside the tower was used to teach ventilation.

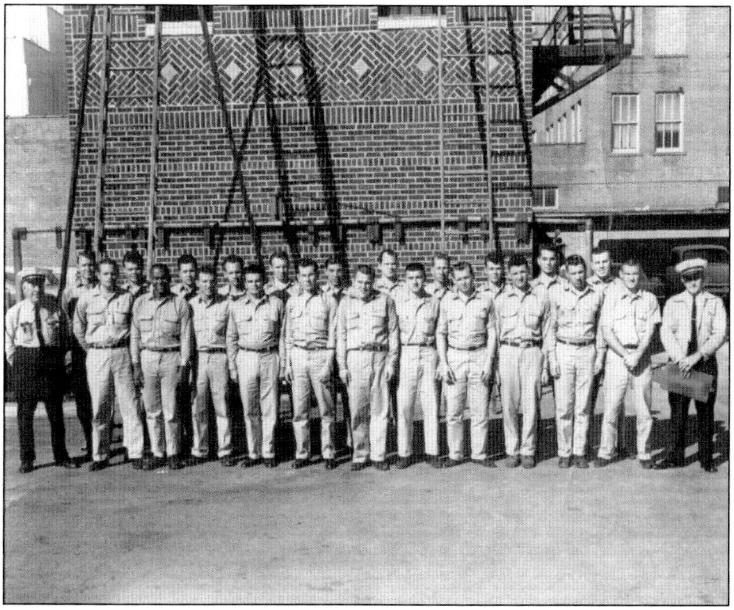

Training officers stand for a group photograph in front of the drill tower. The city's first fire-training facility was opened at the end of 1928 at a cost of $22,000. The facility served the department until 1969, when a new academy complex opened at 8030 Braniff Street, just south of Hobby Airport in far southeast Houston. The new facility cost $1 million.

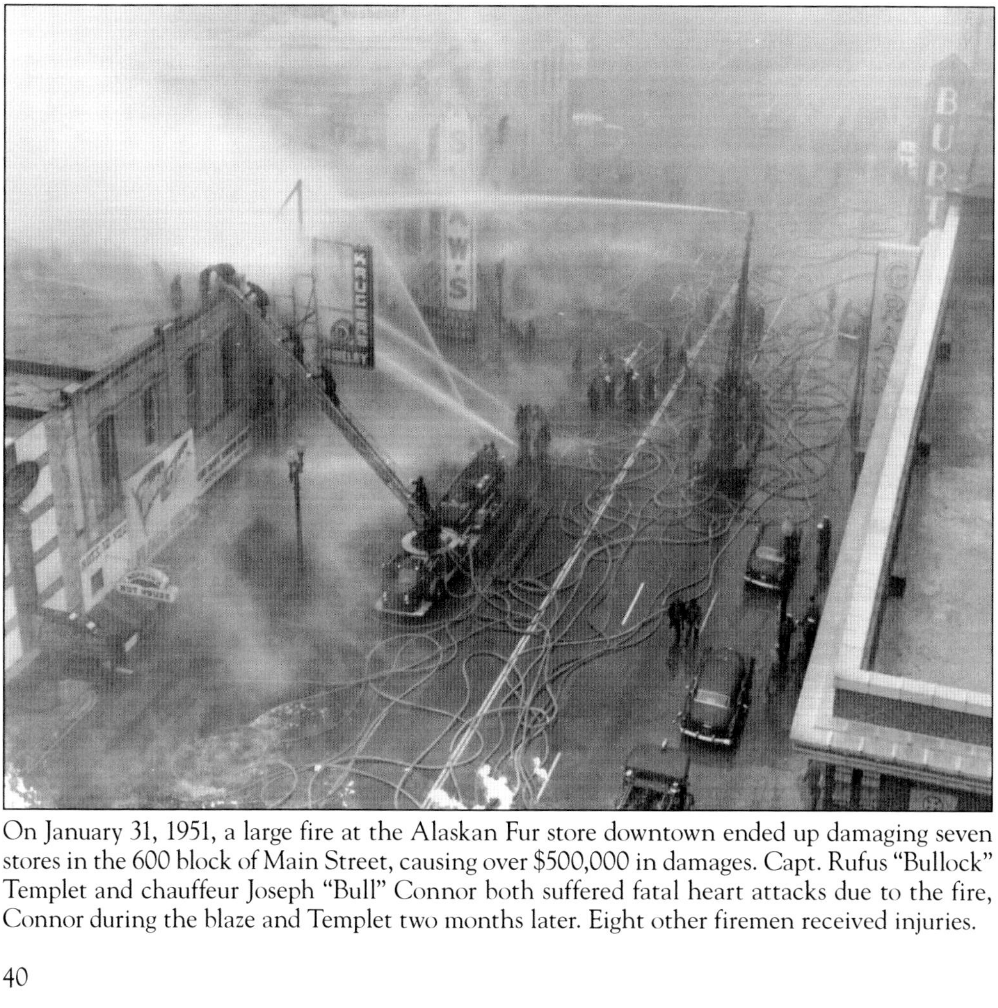

On January 31, 1951, a large fire at the Alaskan Fur store downtown ended up damaging seven stores in the 600 block of Main Street, causing over $500,000 in damages. Capt. Rufus "Bullock" Templet and chauffeur Joseph "Bull" Connor both suffered fatal heart attacks due to the fire, Connor during the blaze and Templet two months later. Eight other firemen received injuries.

On the afternoon of June 5, 1953, a warehouse belonging to the Alco Fireworks Company at 1208 Rosine Street in the Montrose area blew up. Two tons of explosives were detonated, destroying the warehouse and setting several nearby houses on fire. In the explosion, two women and two children were killed. Over 70 people were injured, and property loss was estimated at $3 million.

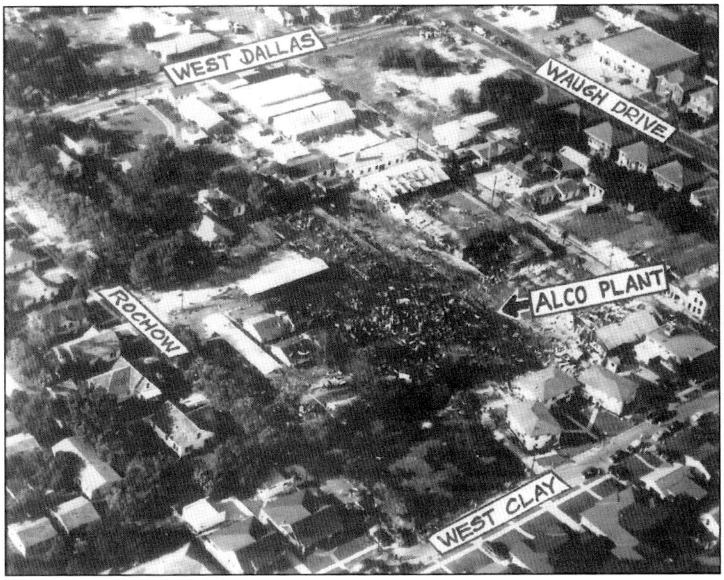

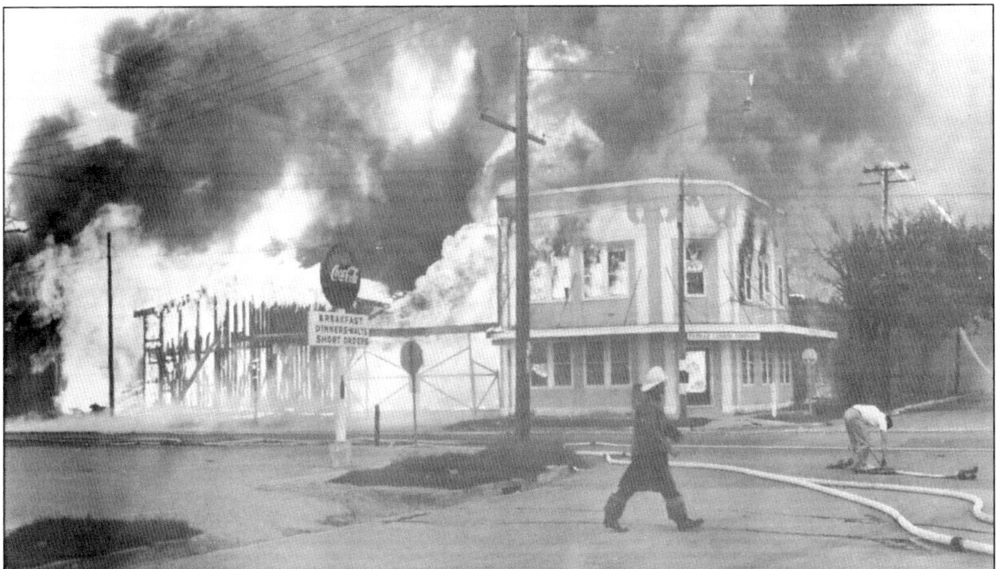

In the fall of 1953, four lumberyards in the southeast part of the city were destroyed by an arsonist. In the early afternoon of September 7, firefighters battled a blaze at the Farrar Lumber Company, which was bounded by Texas Avenue and Dowling, Prairie, and Saint Charles Streets. The business, which covered an entire block, suffered a total loss.

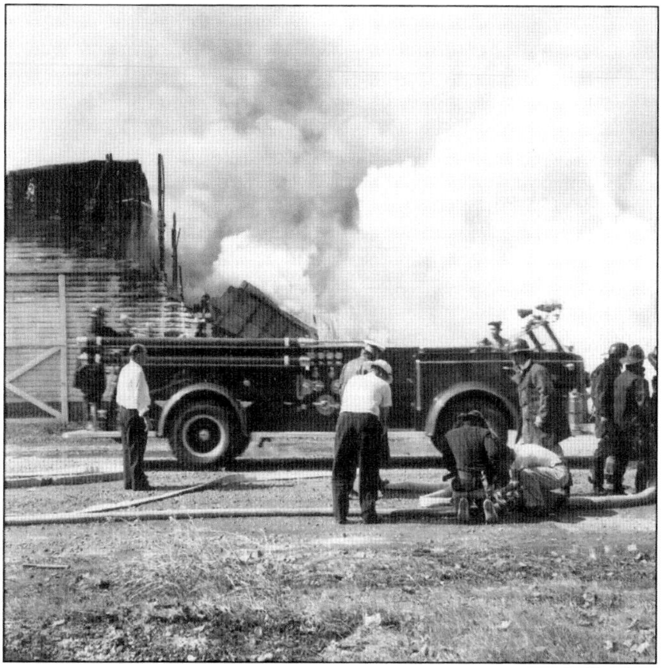

On October 25, the arsonist struck three more lumberyards. The Burton Lumber Company, the SoRelle Lumber Company, and the Winkler Drive Lumber Company were hit. The first was a four-alarm blaze, and the last two were three-alarm fires. Each fire resulted in a $50,000 property loss. On January 5, 1954, an 18-year-old was arrested and charged with arson in all four fires. He was later institutionalized.

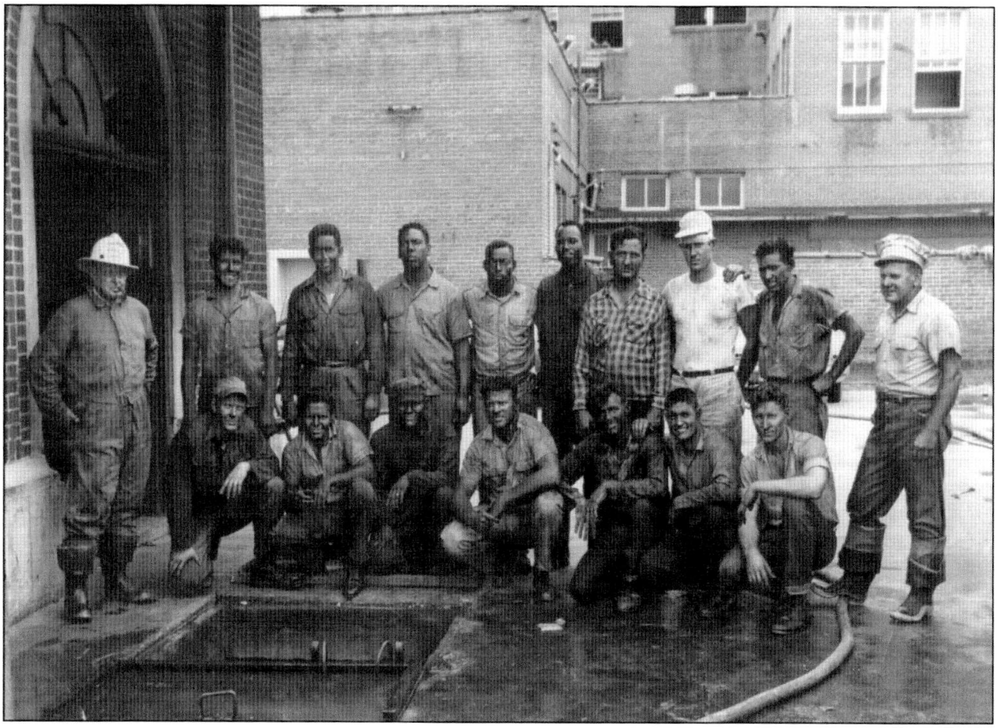

Cadets are seen after smoke school training in this 1955 class photograph. From a recruit's first day on the job, firefighters are consistently involved in training to help them develop the skills to meet whatever challenge lies ahead. Cadets are expected to learn the basics of roughly seven general subject areas: firefighting, rescue, medical, hazardous materials, driver training, ladders, and special services (mass destruction, terrorist attacks, and chemical warfare).

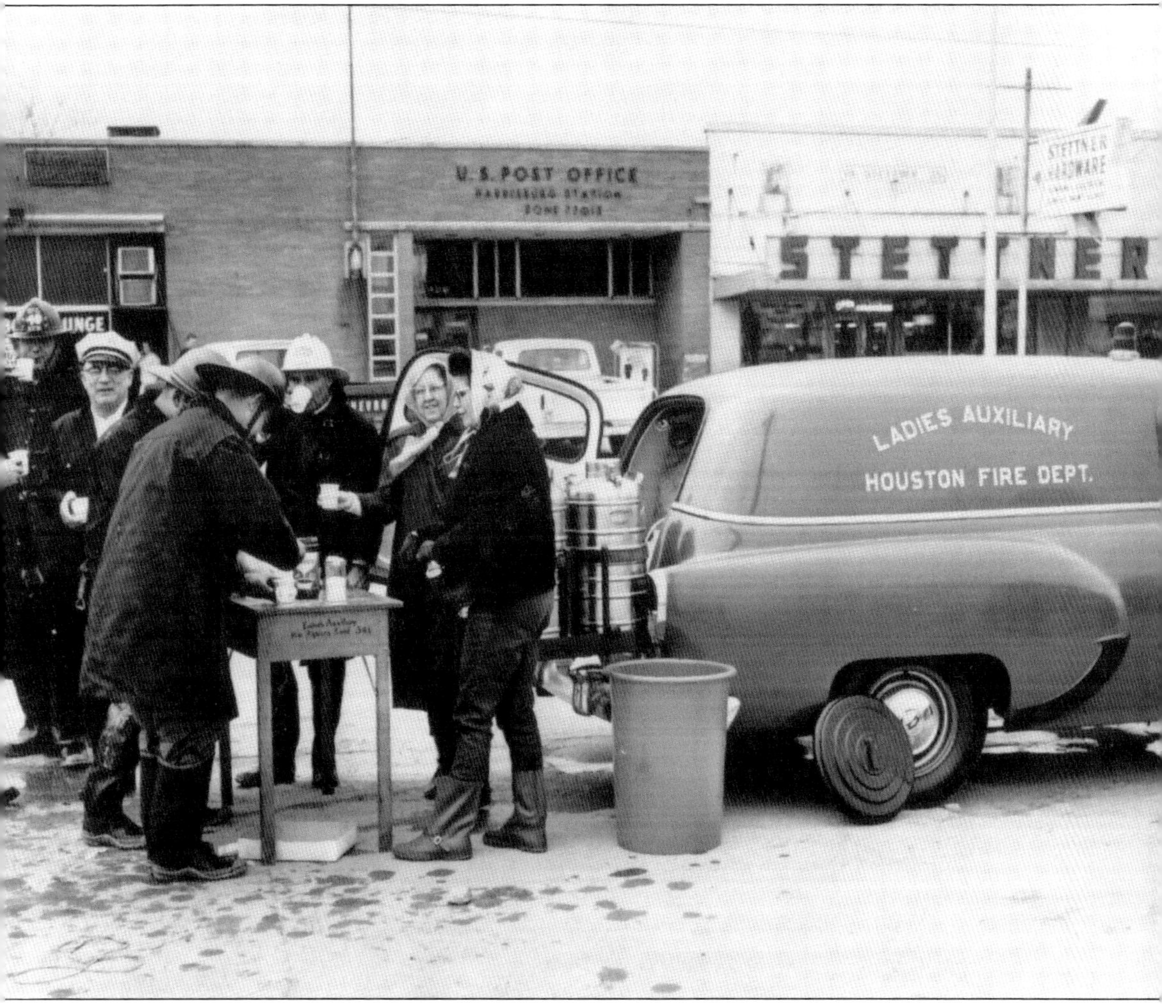

The Ladies Auxiliary began in the early 1950s, growing out of an instance in which Rufus Templet's wife woke in the middle of the night to bring her husband coffee while he fought a fire. This expanded to a large group of wives who became affiliated with the Fire Fighter's Union. In 1954, the auxiliary purchased a Chevrolet station wagon, used to serve coffee and snacks at large fires.

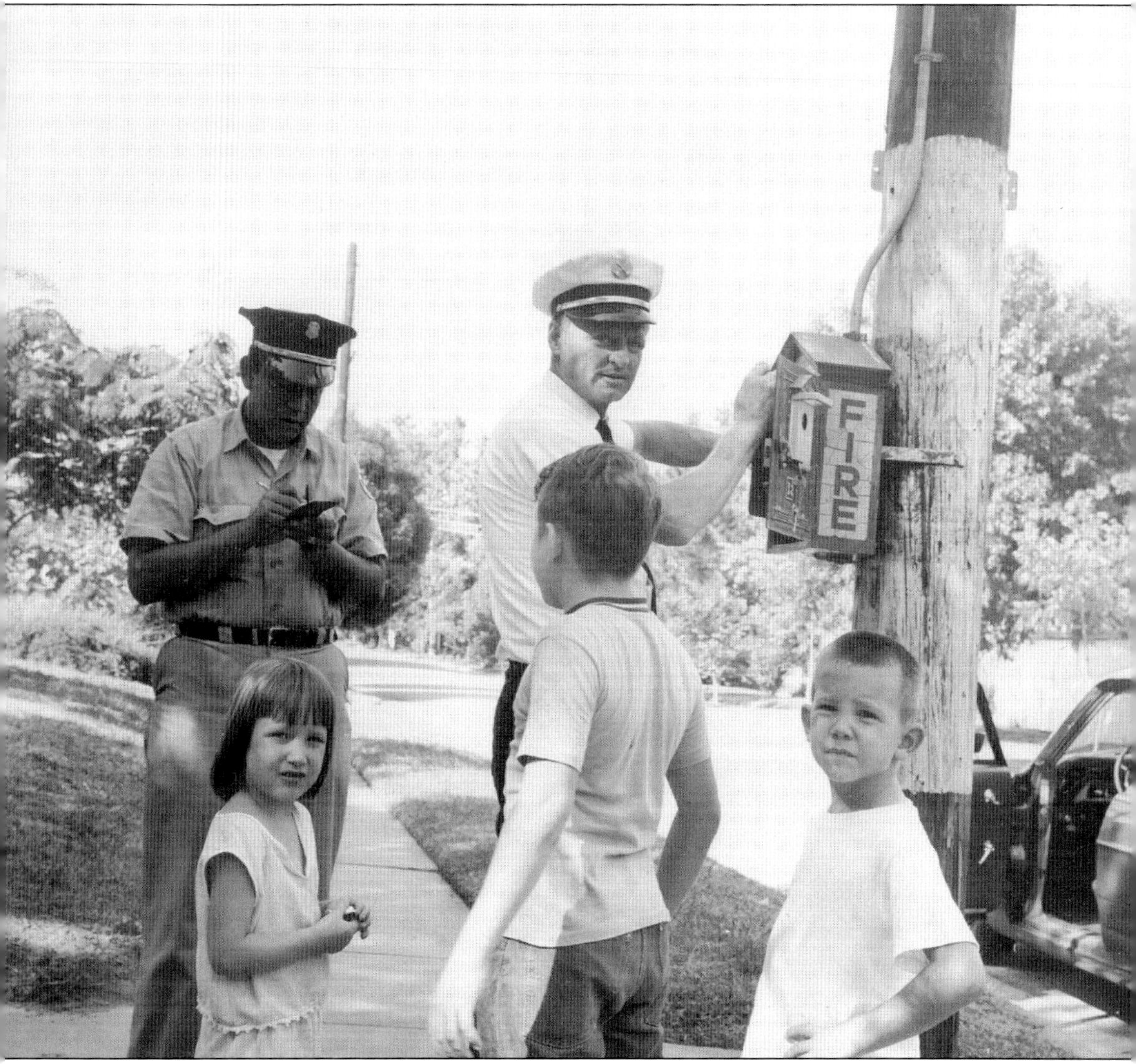

Hundreds of Gamewell telegraph boxes provided Houstonians with the means of turning in fire alarms from 1892 until the 1970s, when the department adopted a new system designed for keeping fire records on a computer. Of the 3,748 alarms received by the boxes in 1975, a total of 3,430 were false. By the spring of 1977, only about 100 boxes remained, mostly in the downtown area.

Three

A One-Horse Town Grows

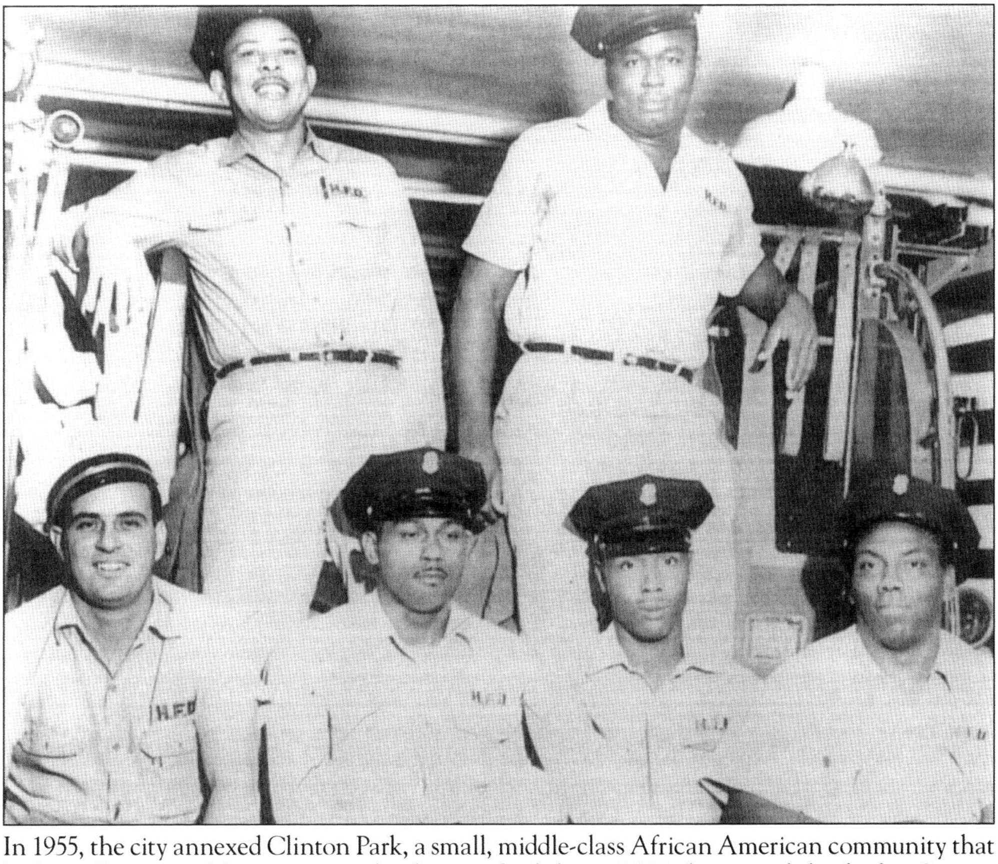

In 1955, the city annexed Clinton Park, a small, middle-class African American community that had a well-equipped fire station and volunteer firefighters. HFD designated the facility Station 42. That August, nine black men graduated from the training academy, becoming the city's first black firefighters. Later, Rudolph Cline became the HFDs first black district chief, and Otis Latin became the first black chief.

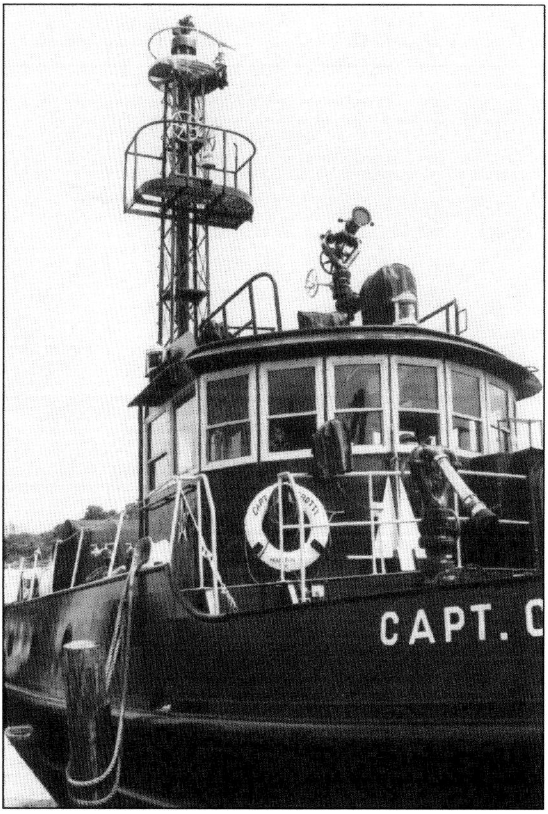

On September 1, 1950, the Port of Houston's second fireboat went into service. The *Captain Crotty* replaced the 25-year-old *Port Houston*, ending the city's contract with the Navigation District, which took control of dockside fire protection. Built in Camden, New Jersey, the fireboat served on the channel for 33 years until it was retired in 1983.

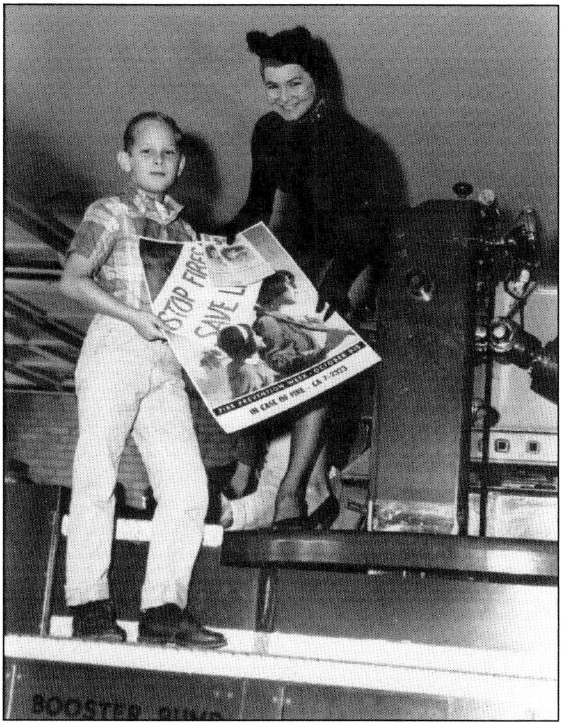

Houston had its own television star for kids before those on PBS. Bunny Orsay, better known as KiTiRiK, was the black cat mascot for KTRK, channel 13. KiTiRiK, a fixture in Houston, made numerous appearances in public and supported many causes and contests involving children, including the Fire Safety Poster Contest for Houston students, a competition that continues to this day.

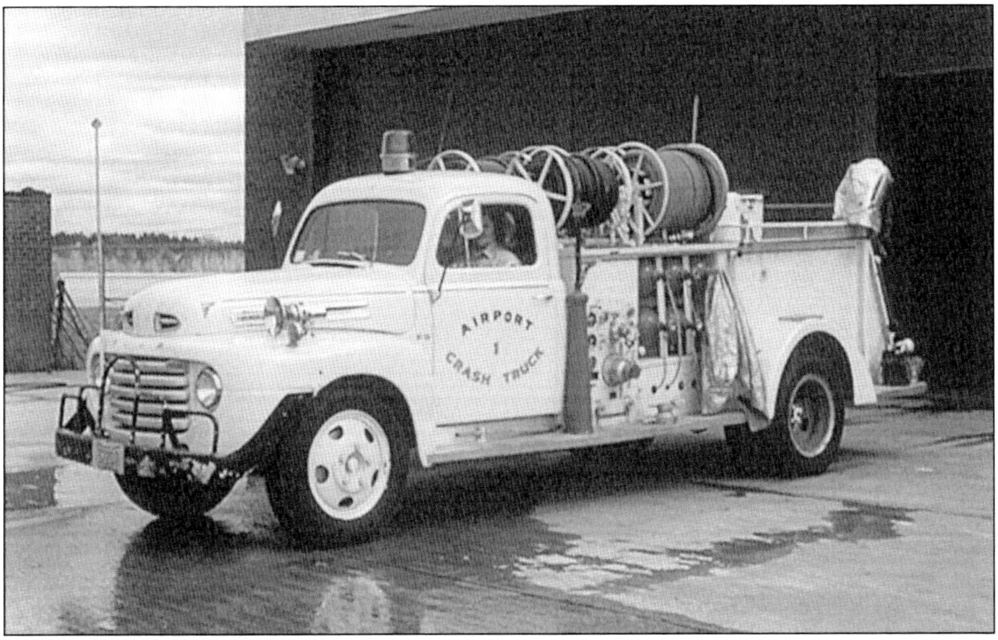

Originally, Station 36's home was a World War II–era wooden structure next to Houston Municipal Airport's terminal building. A new station opened in 1957 at the northern edge of the airport. Crash-rescue apparatus, like this 1949 Ford Darly CO2, operated out of the back of the station. Expansion of the airport in the 1980s caused a relocation of the crash trucks inside the airport.

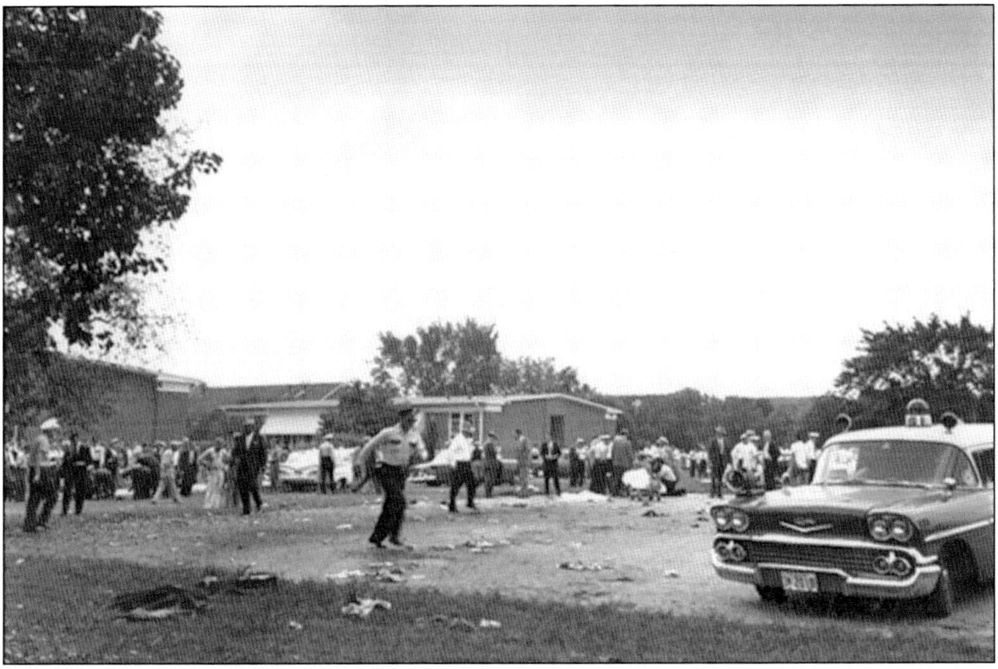
On September 15, 1959, seven people were killed on the playground at Poe Elementary School on Hazard Street at North Boulevard when a man distraught over his marital problems detonated a suitcase full of explosives. Two school employees and four children died, including the man's son. In addition, 16 children were injured.

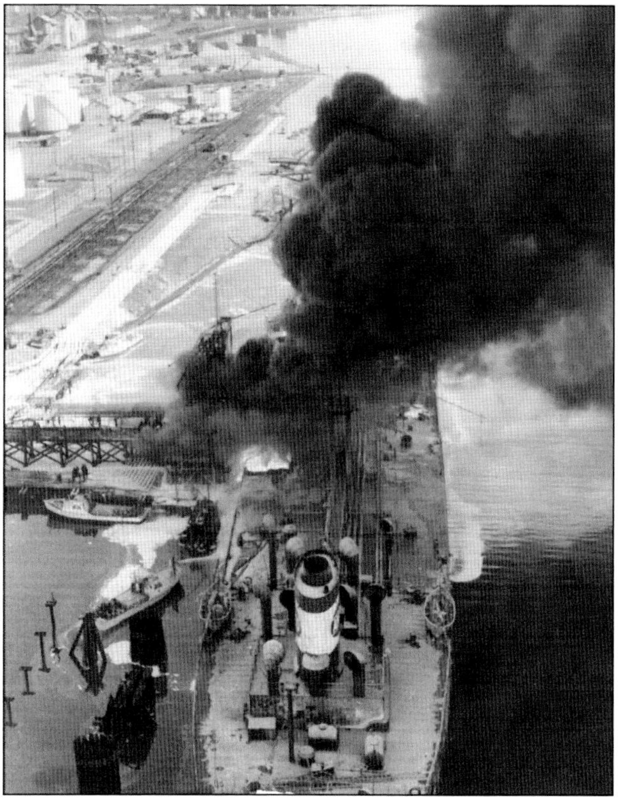

Shortly after midnight on November 18, 1959, there were two explosions aboard the docked *Amoco Virginia*. Within minutes, the port side of the ship and much of the superstructure were in flames. It took 500 firemen from Houston and surrounding departments 16 hours to put the fire out. Seven crew members and firefighter D.H. Chandler died; 25 crew members and 13 firefighters were injured.

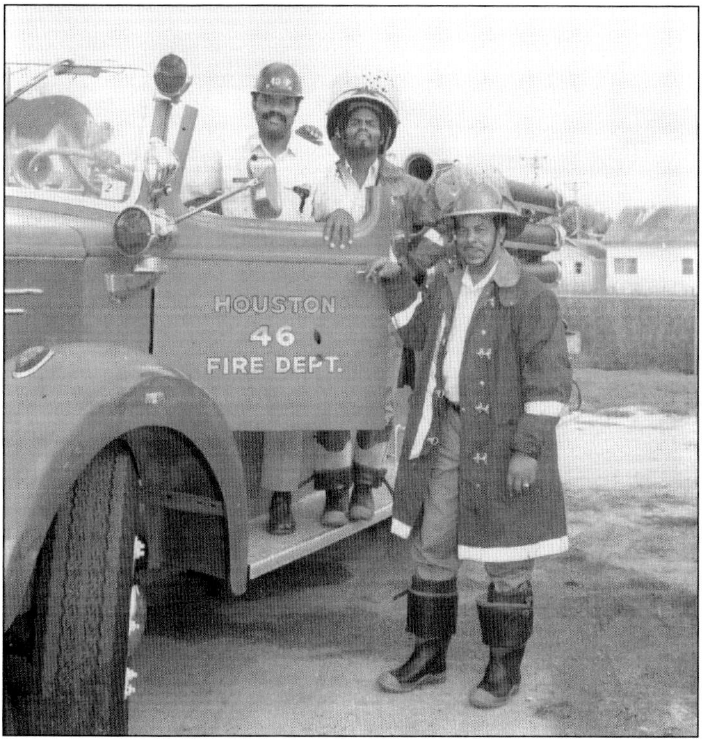

Fire Station 46 opened in 1956 in a former volunteer station at Corinth and Stassen Streets in southeast Houston. The city had doubled its land area for the second time in seven years; it annexed 27 outlying districts, adding 184 square miles. In 1976, a new station was built on the other side of Loop 610 at Corder and Scott Streets in a single-story four-bay structure.

The department's Arson Bureau has a long tradition. Its job is to lessen the occurrence of illegal fires through aggressive investigation and prosecution. In the cases involving the line of duty deaths of Capt. Grady Burke and firefighters Kim Smith, Lewis Mayo, and Kevin Kulow, all suspects received lengthy prison sentences. In 1977, Chief Lonnie Franklin died while responding to a fire later determined to be arson.

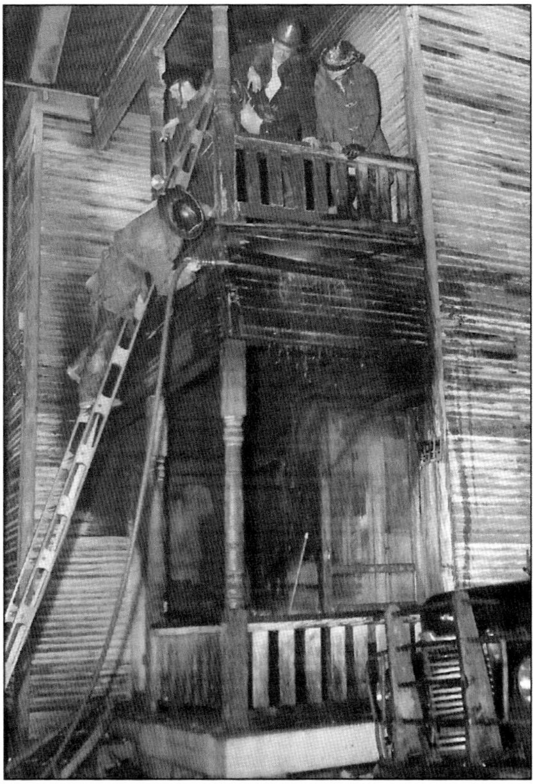

The HFD's light truck was used from 1950 until 1977. The vehicle was automatically dispatched on all fires of two alarms or above that happened after sunset. The chassis is a 1950 Ford truck, and the body was built at the department's repair shop by Frank Viola.

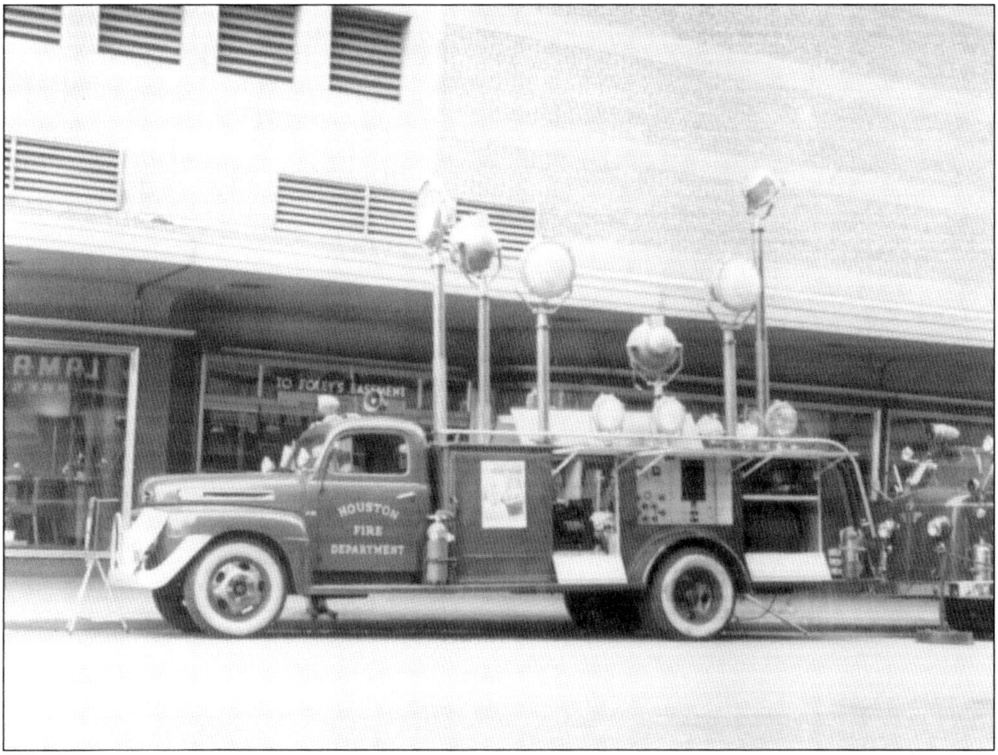

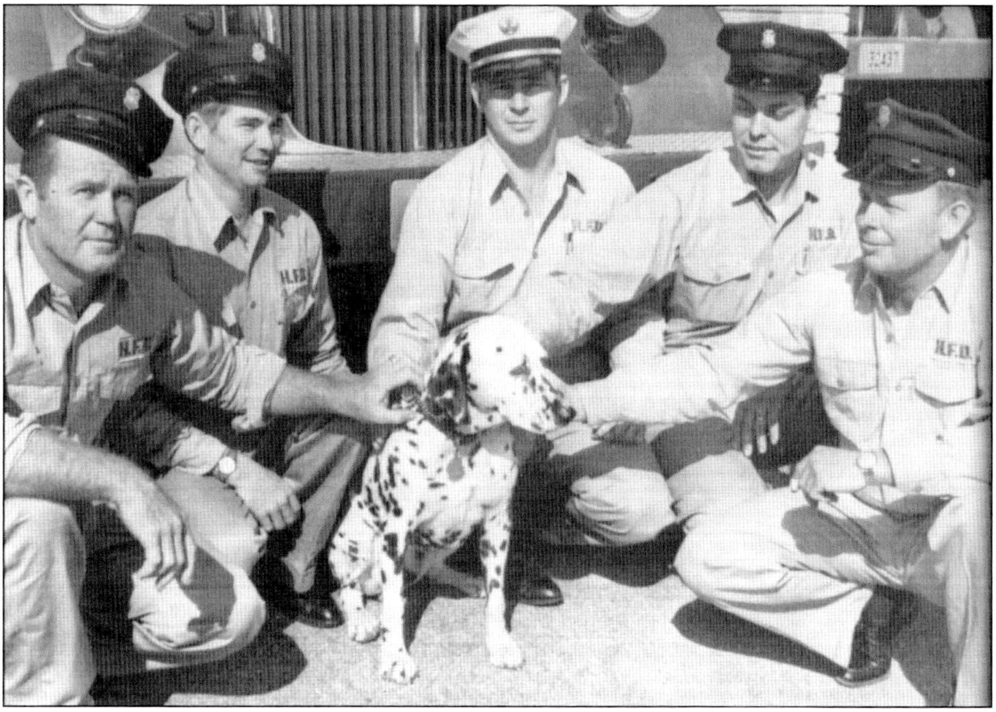

The crew of Station 28 poses in 1962 with station mascot Domino. Firehouses in Houston and throughout the world have had station pets or mascots, with the Dalmatian becoming synonymous with the fire service. The tradition stems from the horse-drawn era, when Dalmatians served as companions and guards for the horses and firefighters.

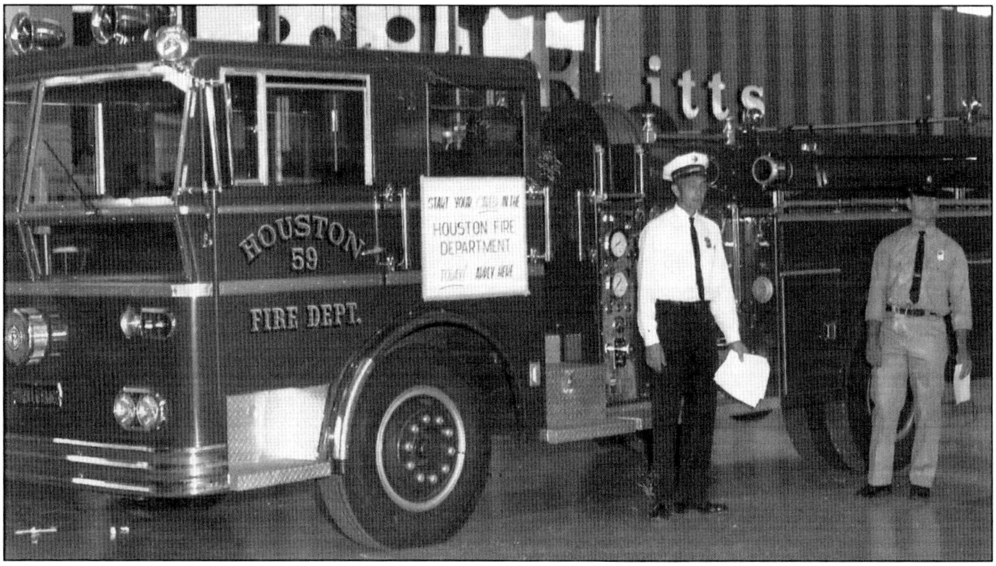

The fire department serves the public in numerous ways, including outreach and community education. The goal of such programming is to perform fire-safety education, make firefighters more visible to the community, and facilitate effective communication inside and outside the department. Seen here in the 1960s, two HFD officials pose with an apparatus from Station 59 outside Northline Mall's Britt's Department Store.

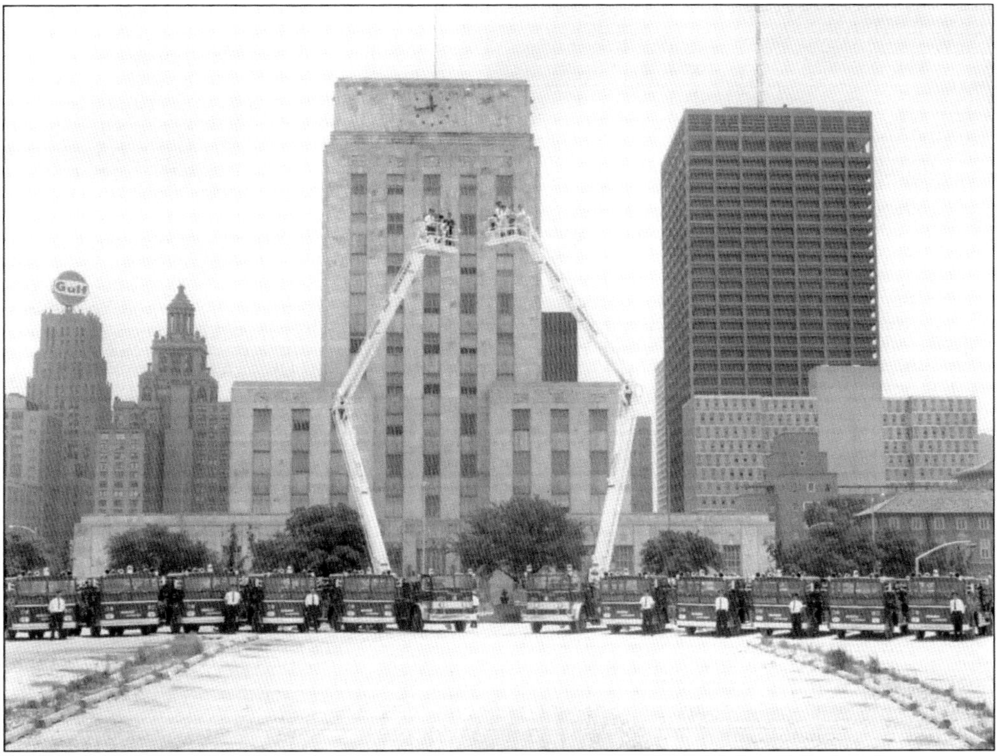

New equipment is displayed in 1966. The department bought several Ward LaFrance trucks to replace its aging fleet of engines. Typically, new apparatus would arrive by rail. Soon thereafter, the department would line the new apparatus up for a photo shoot to show off the new additions to its fleet. Here, the space in front of City Hall has been chosen as a backdrop.

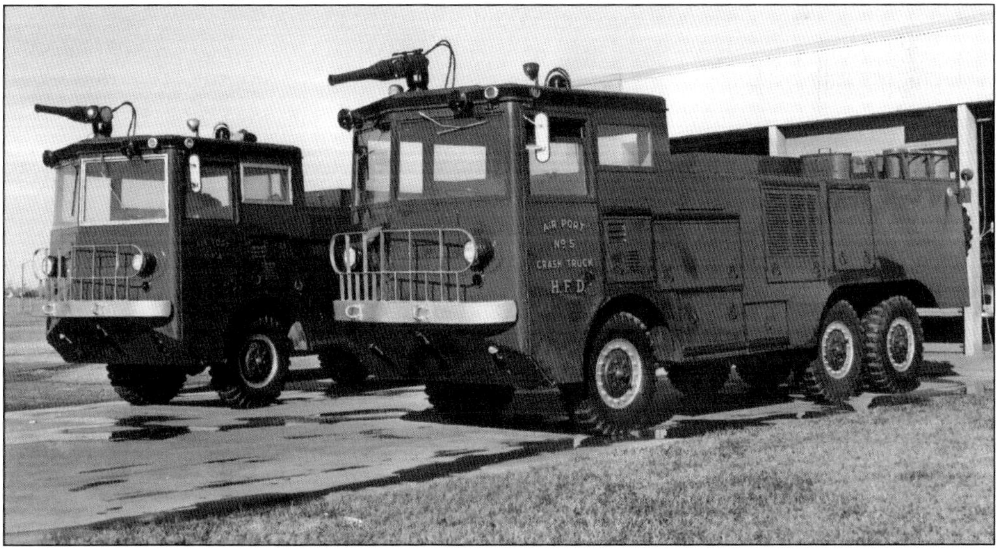

Fire Station 54, originally housed in the old Almeda Plaza community hall and fire station, was relocated to a temporary wooden structure at Intercontinental Airport in 1965 as Houston's new airport neared its opening. A permanent Station 54 was completed in 1969 at the end of two runways. It housed several crash-rescue trucks, like the 1954 American LaFrance O-10s shown here.

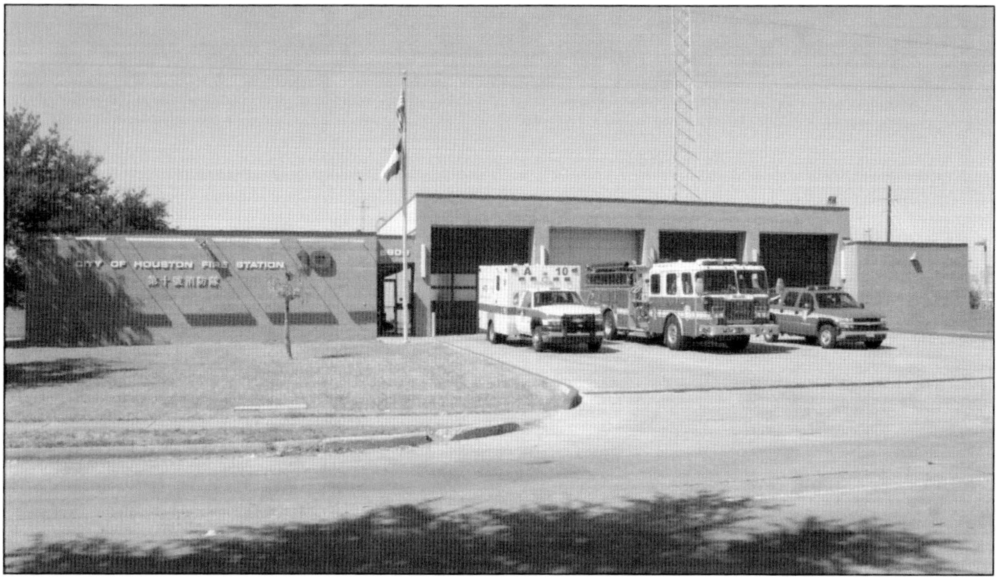

The crew of Station 10 was housed in three buildings before this facility, located at 6600 Corporate Drive at Clarewood Drive in west Houston, was moved in 1985. Fire Station 10 at Bellaire Boulevard, and Station 76, Alief Community, both serve Houston's new Chinatown district. Station 10 was previously located in East Downtown near the George R. Brown Convention Center, an area now known as Old Chinatown.

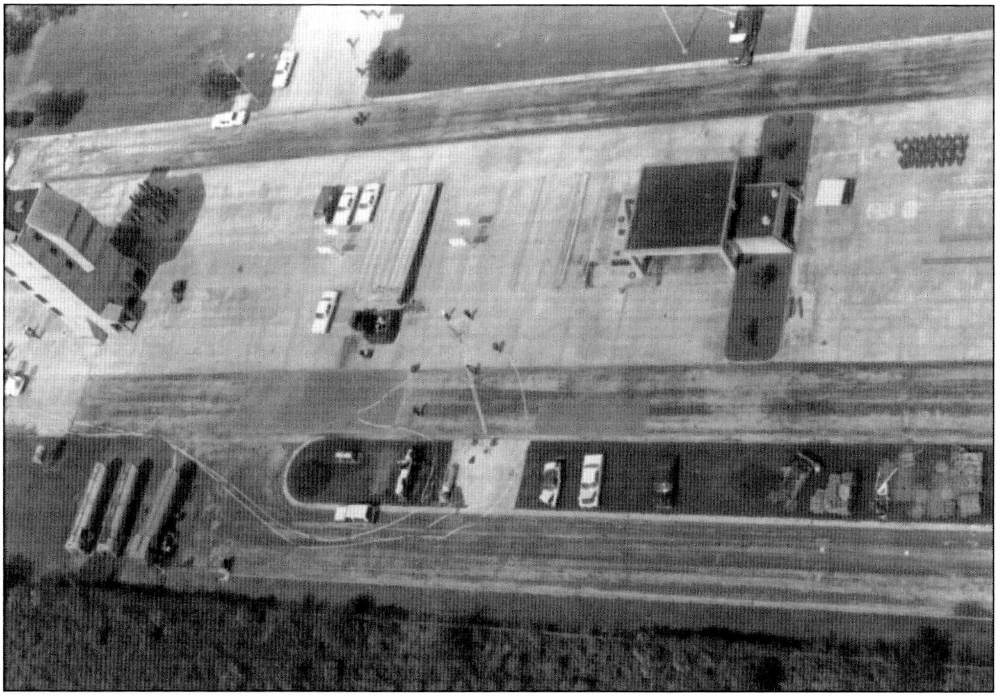

This aerial photograph shows the Houston Fire Department Training Academy, known now as the Val Jahnke Training Facility. It opened in November 1969. Located next to Hobby Airport, this modern facility replaced the old academy that had been built in 1928, across the street from the old Central Station.

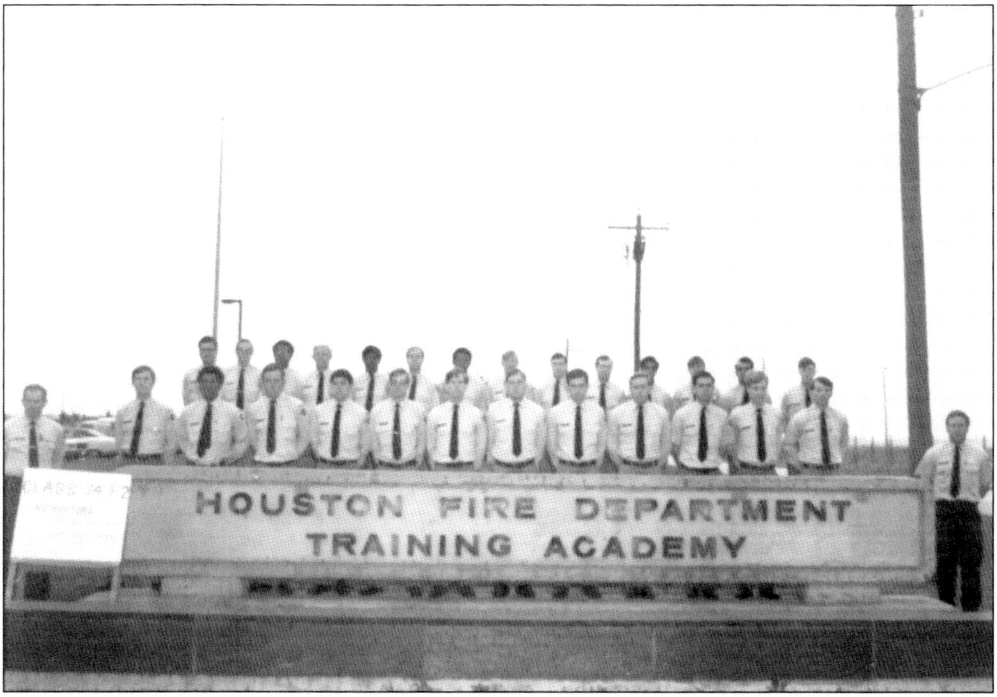

The HFD Training Academy facility was constructed in 1969 with two classrooms, a service garage, a multipurpose auditorium, and an office building. At the rear of the 15-acre site were a six-story drill tower and a two-story fire building for recruit training. Hundreds of HFD cadets train at the academy to become part of one of the largest fire departments in the United States.

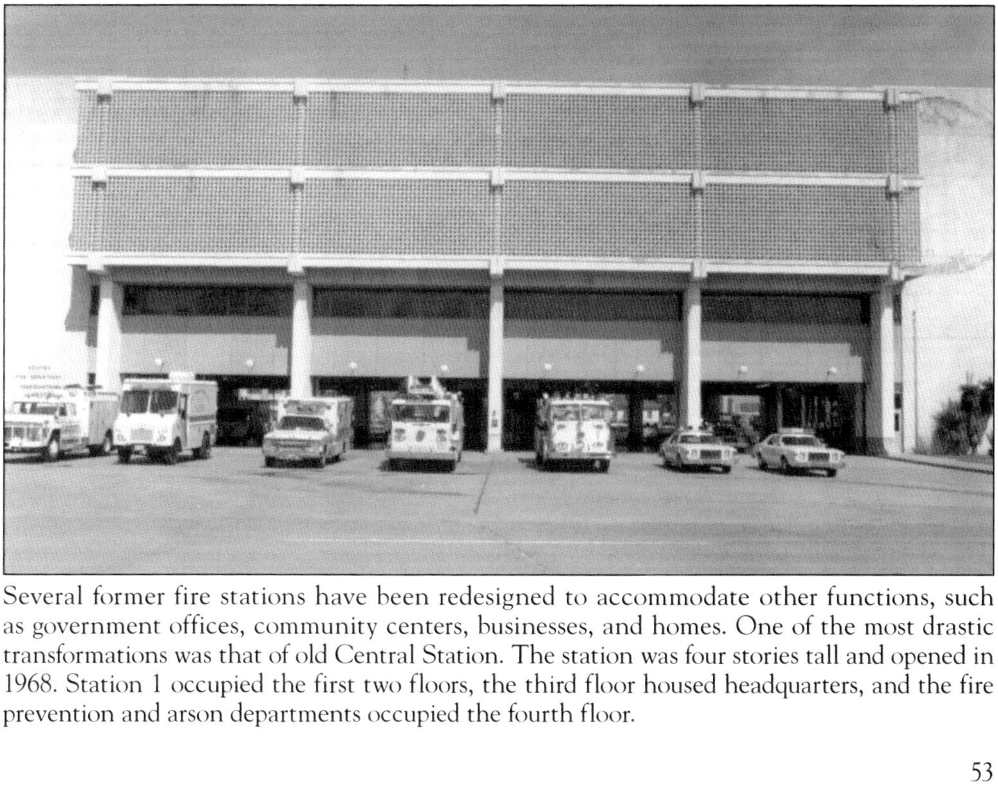

Several former fire stations have been redesigned to accommodate other functions, such as government offices, community centers, businesses, and homes. One of the most drastic transformations was that of old Central Station. The station was four stories tall and opened in 1968. Station 1 occupied the first two floors, the third floor housed headquarters, and the fire prevention and arson departments occupied the fourth floor.

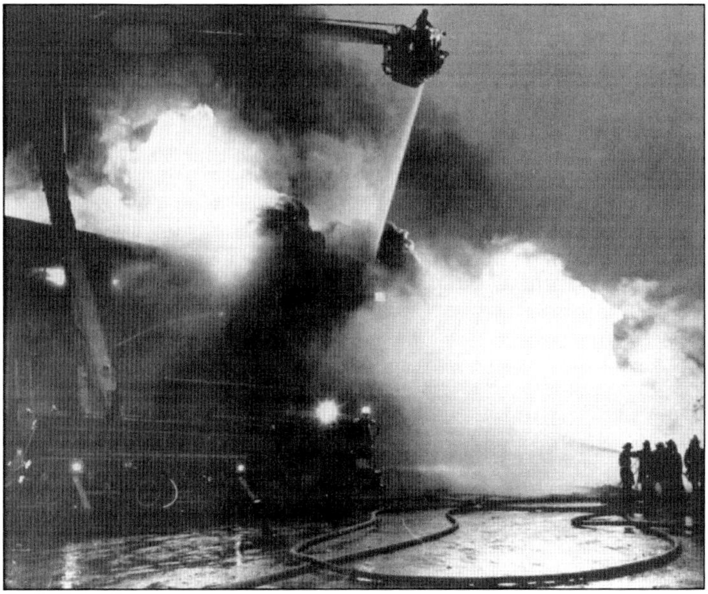

An HFD snorkel delivers a heavy stream onto the burning Lack's Wholesale Distributors warehouse at 1315 Baker Street on January 10, 1970. This fire destroyed the block-long warehouse and demanded the largest response of fire trucks in the city's history at that time. During the four-hour battle, twenty-nine engines and seven ladders were used to bring the fire under control. The loss was estimated at $1.5 million.

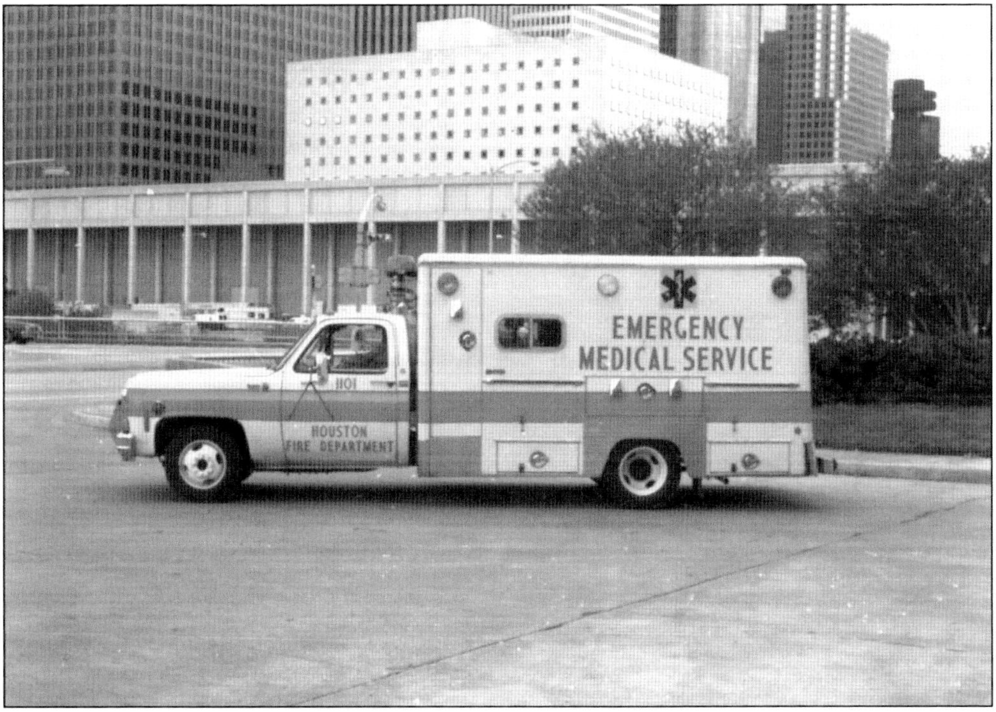

A study done by members of local medical and public-service communities concluded that Houston's emergency medical service would be handled best by the city's fire department. HFD already had housing, crews, a maintenance facility, and a highly efficient dispatch system. In addition, the fire department had achieved short response times, and its members had experience. The HFD began providing emergency medical service in April 1970 under the direction of District Chief L.O. "Whitey" Martin.

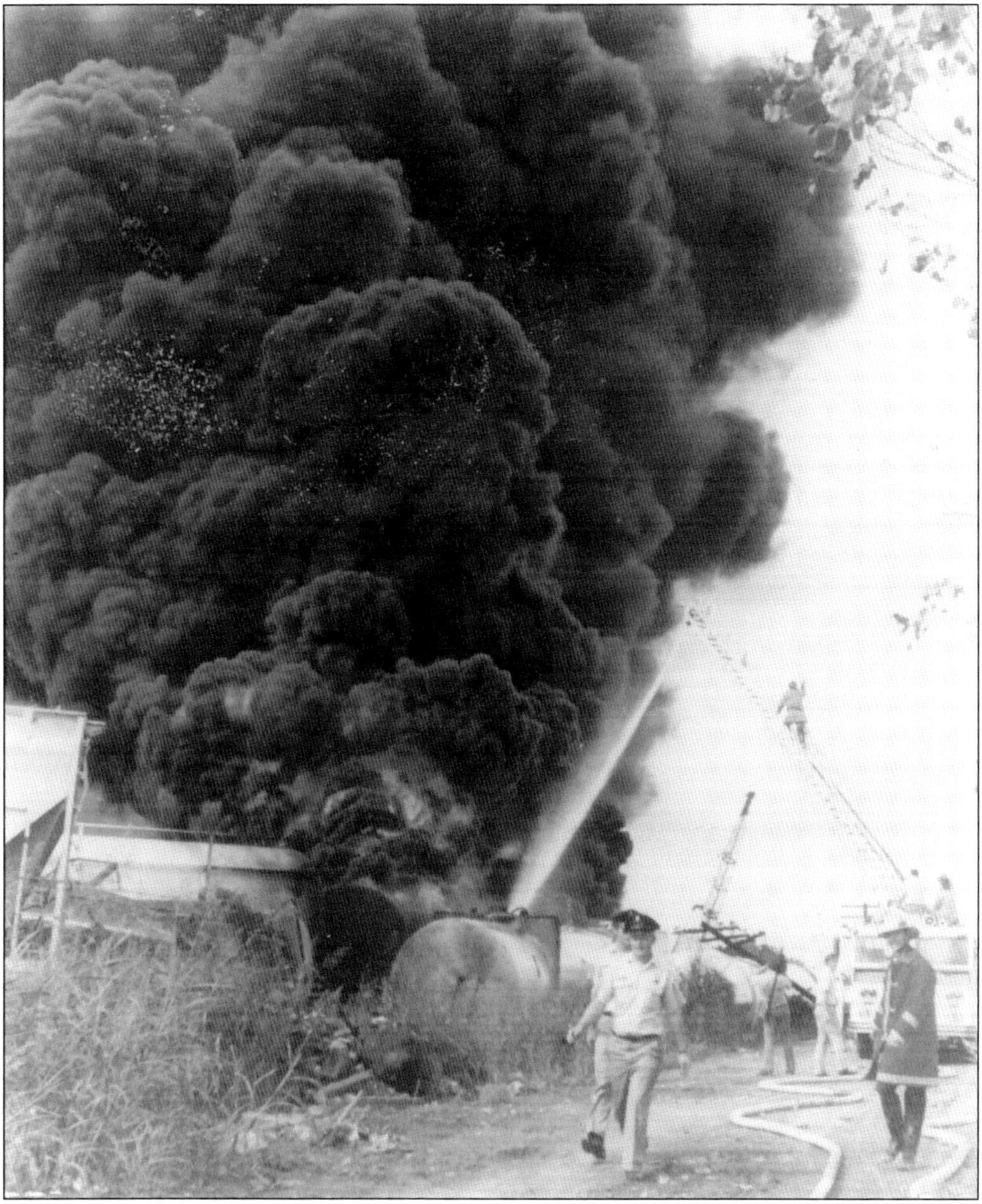

On the afternoon of Tuesday, October 19, 1971, a northbound Missouri Pacific Railroad train comprising 16 cars derailed along Mykawa and Almeda-Genoa Roads. The track had buckled under the weight of the loaded railcars. One tank car, laden with highly flammable butadiene, ruptured, caught fire, and, after being exposed to the flames for over an hour, exploded.

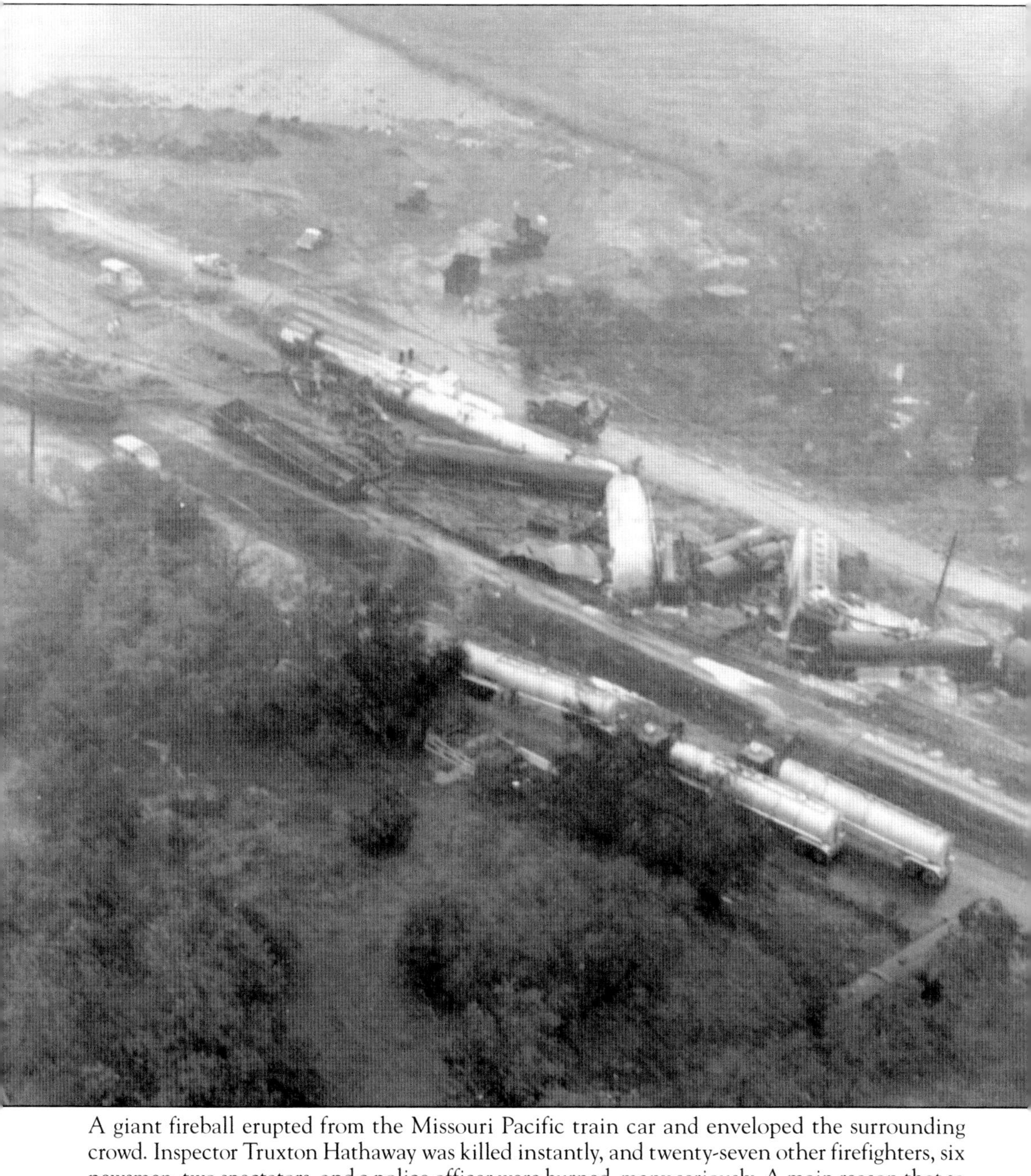

A giant fireball erupted from the Missouri Pacific train car and enveloped the surrounding crowd. Inspector Truxton Hathaway was killed instantly, and twenty-seven other firefighters, six newsmen, two spectators, and a police officer were burned, many seriously. A main reason that so many were hurt was that fire officials did not have information on the chemicals in the railcars, making it hard for them to handle the situation properly.

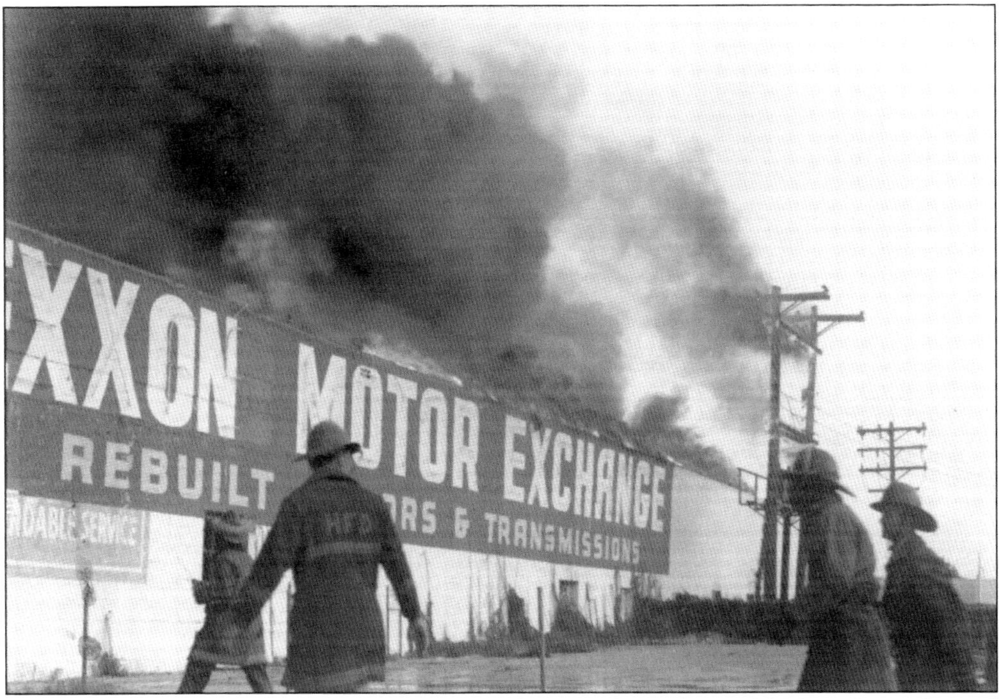

The Texxon Motor Exchange, an engine-repair business located at 2400 Main Street, burned down on October 7, 1973. The four-alarm fire damaged over 40 cars and vans that were parked inside the one-story structure when the blaze started. Business owner Jack Enright estimated the loss at over $400,000.

A railcar leaking liquefied petroleum gas exploded and started a rapidly spreading fire on September 21, 1974, at Southern Pacific's Englewood Yards at Homestead and Liberty Roads. The leak came from a car damaged when it was coupled with a switch engine. The blast leveled several nearby metal warehouses, caused the evacuation of 1,200 people, and caused minor explosions in other tank cars.

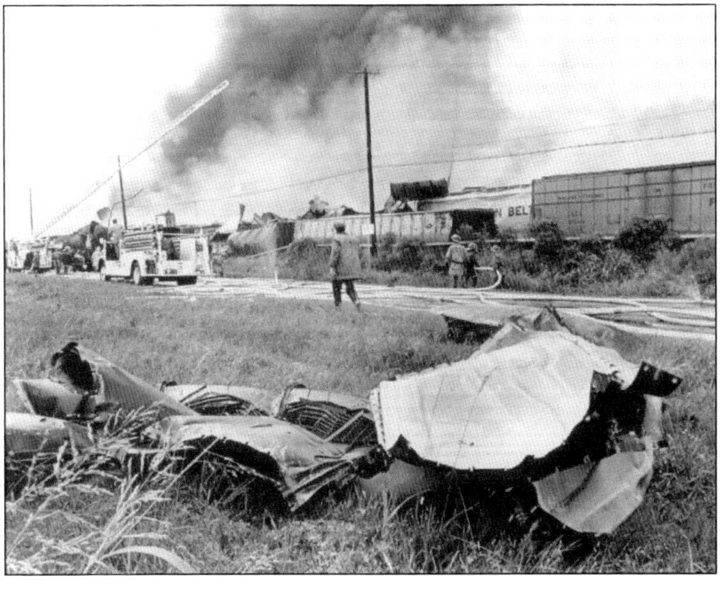

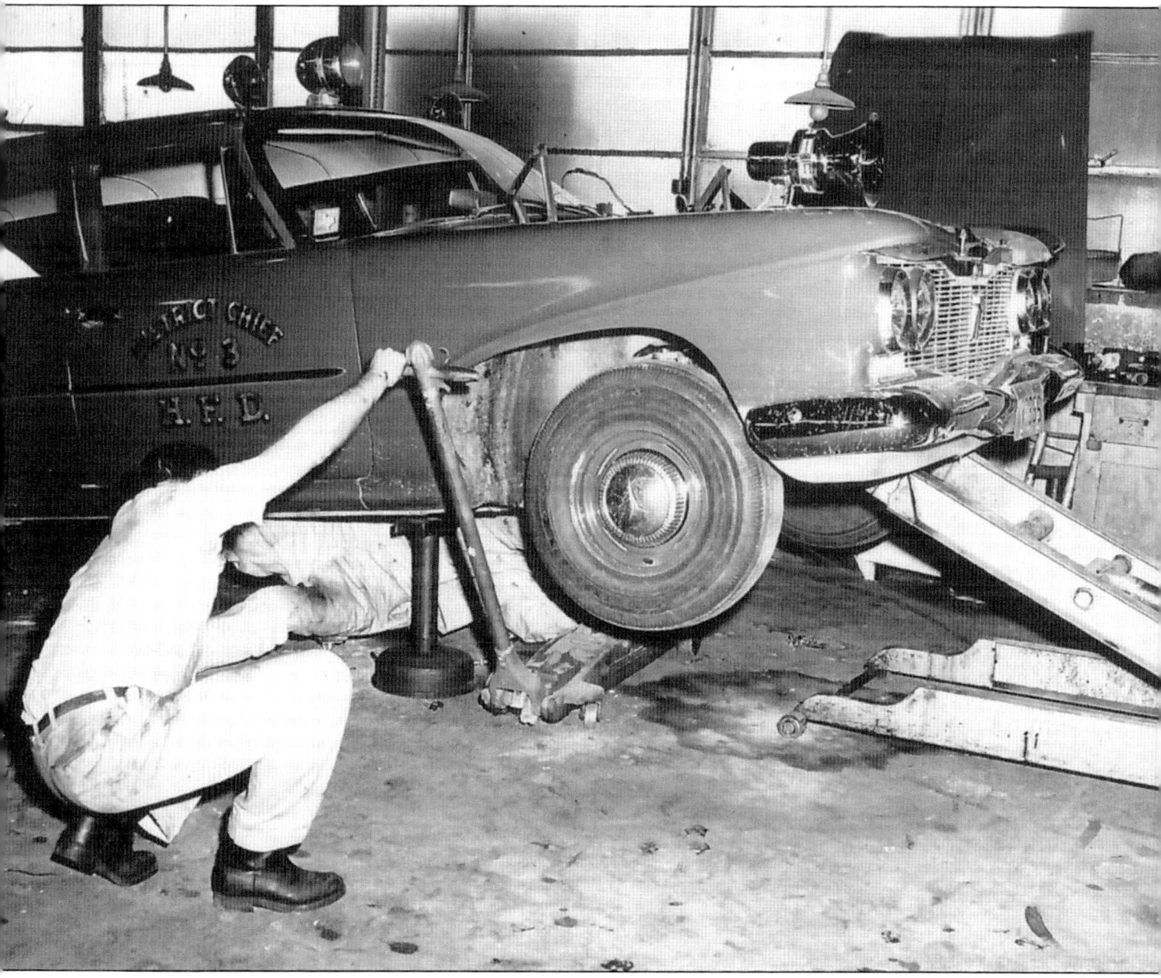

The Fleet Maintenance Division has served the city from three locations. The first opened in 1924 at Freeman and Hogan Streets, moved to Girard and Elder Streets in 1970, and finally settled in at the department's administrative headquarters at 1205 Dart Street. The division has been responsible for the maintenance of over 1,000 wheeled, motorized, towed, and other types of equipment located at over 90 HFD sites throughout the city.

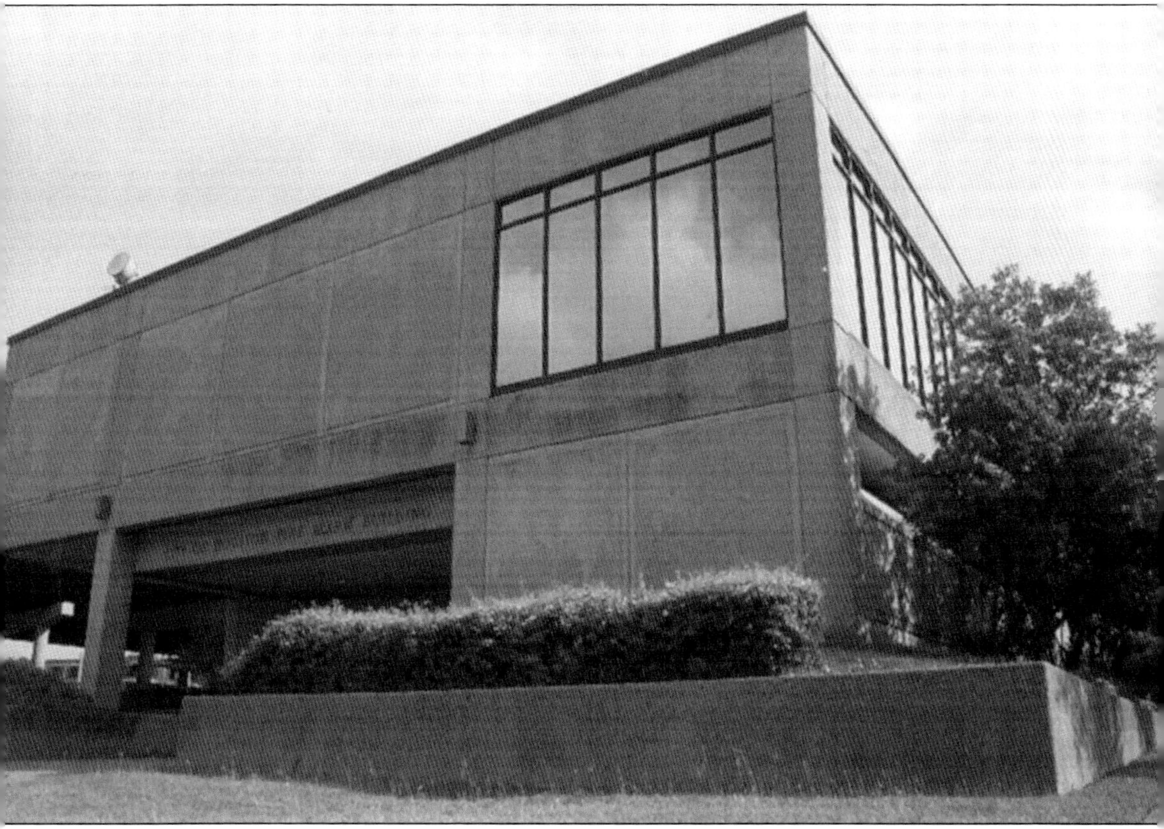

After several delays, the new fire alarm building opened in December 1975. Located at Preston and Bagby Streets, on the block north of Station 1, the new facility housed the dispatching operations center and additional offices. Dispatchers here would see innovations such as the 911 emergency telephone system and the Uniform Fire Information Reporting System, designed in 1976 for keeping computerized fire records.

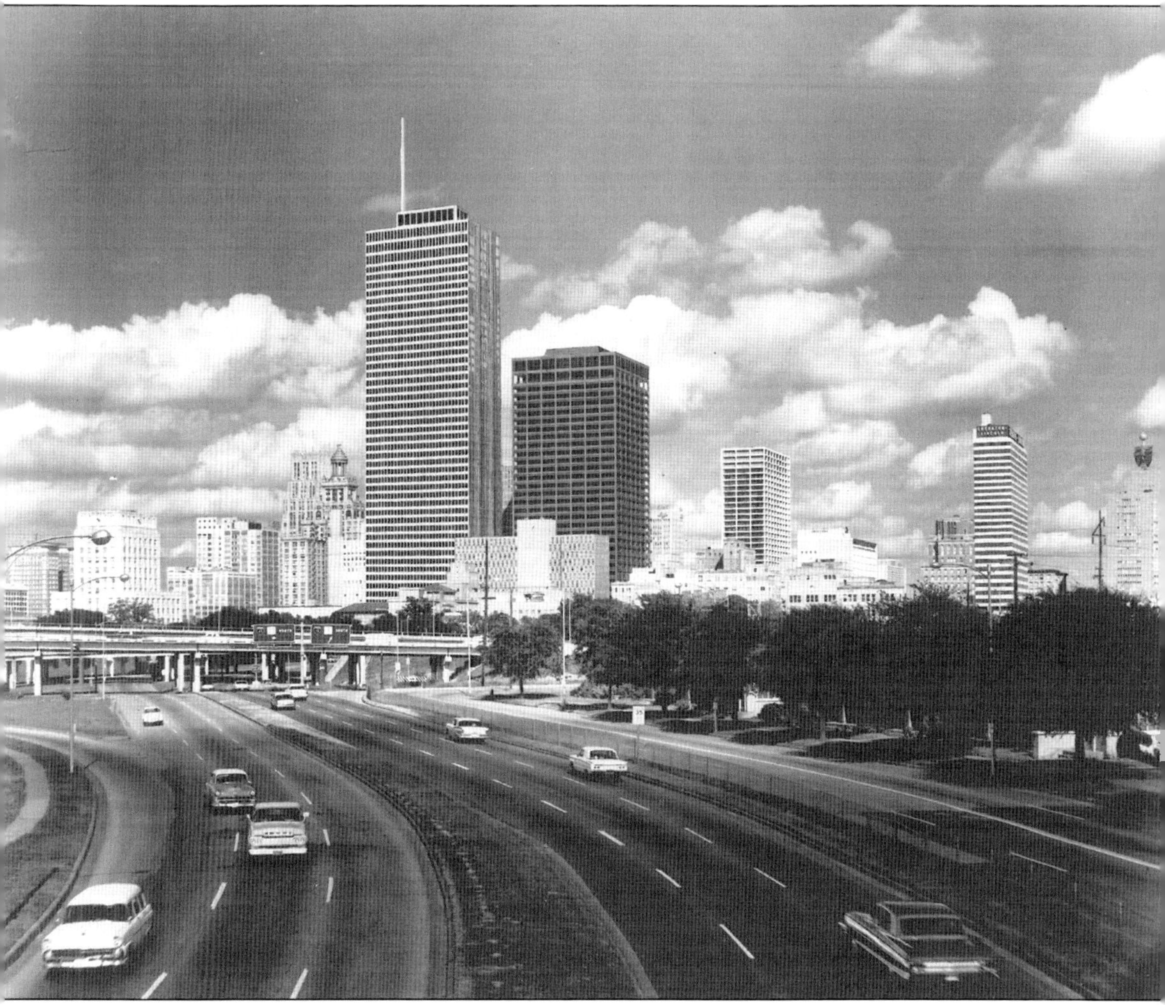

When the new fire alarm building opened in 1975, the department's radio transmitter was placed atop the tallest building in the city, One Shell Plaza. Satellite repeater stations were installed throughout the Houston area. The lights on the fire alarm building's status board could be changed not only by the dispatchers, but also by the firemen in the stations.

Four

BOOM, BUST, AND RECOVERY

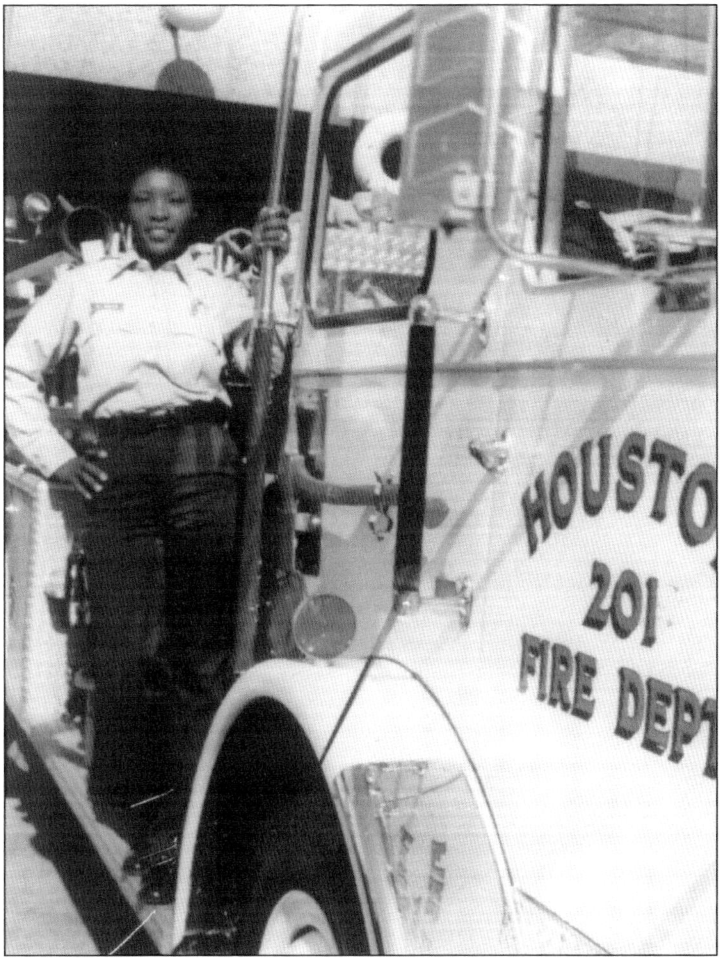

Arandas Madison (pictured) was hired in 1982. She became the HFD's second female firefighter and its first black woman. In 1973, Linda Honeycutt had become the first female firefighter in the department. Honeycutt completed the entrance requirements and the rigorous fire academy, paving the way for other women in the service.

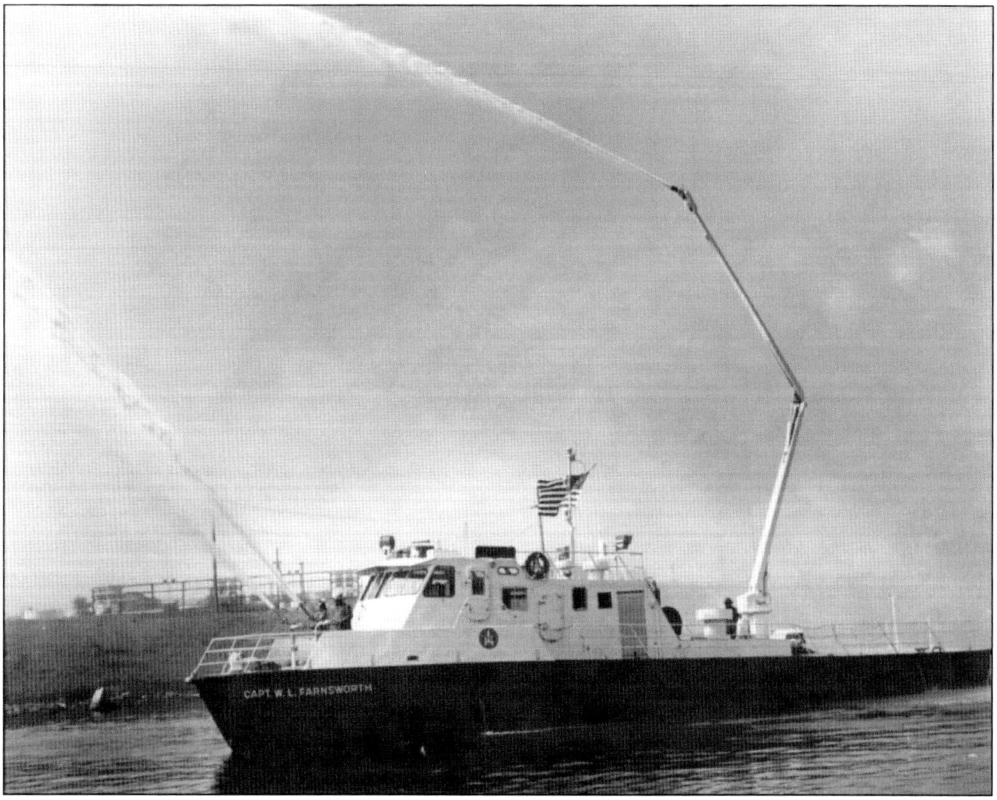

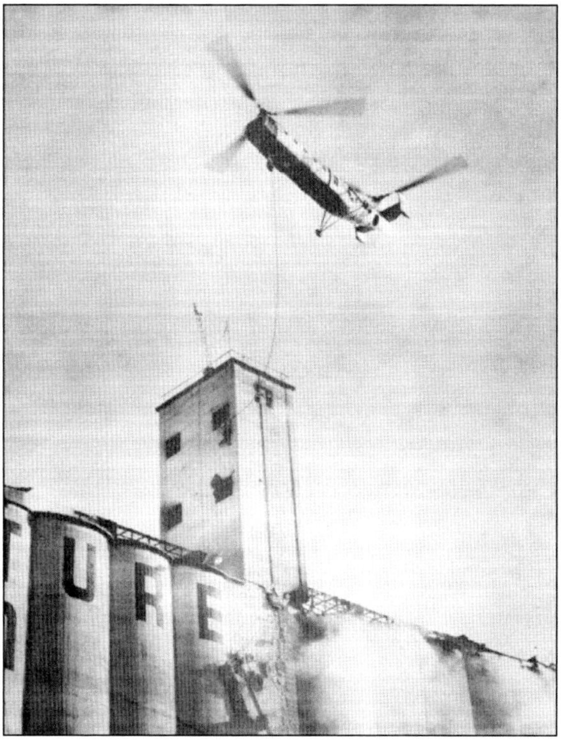

The third fireboat was the *Captain W.L. Farnsworth*, added by the Navigation District (at this time, known as The Port of Houston Authority) to supplement the *Captain Crotty*. The steel boat cost nearly $425,000 and was built in Houston in 1974. The vessel pumped 6,000 gallons of water per minute and was docked along Greens Bayou, just north of the channel.

A quiet Sunday afternoon in east Houston was shattered by a huge explosion at the Goodpasture grain elevator on February 22, 1976. As a result, nine persons at the elevator were killed, and the facility was destroyed, a $42 million loss. Tons of debris made a complete search of the Goodpasture facility impossible until a later date.

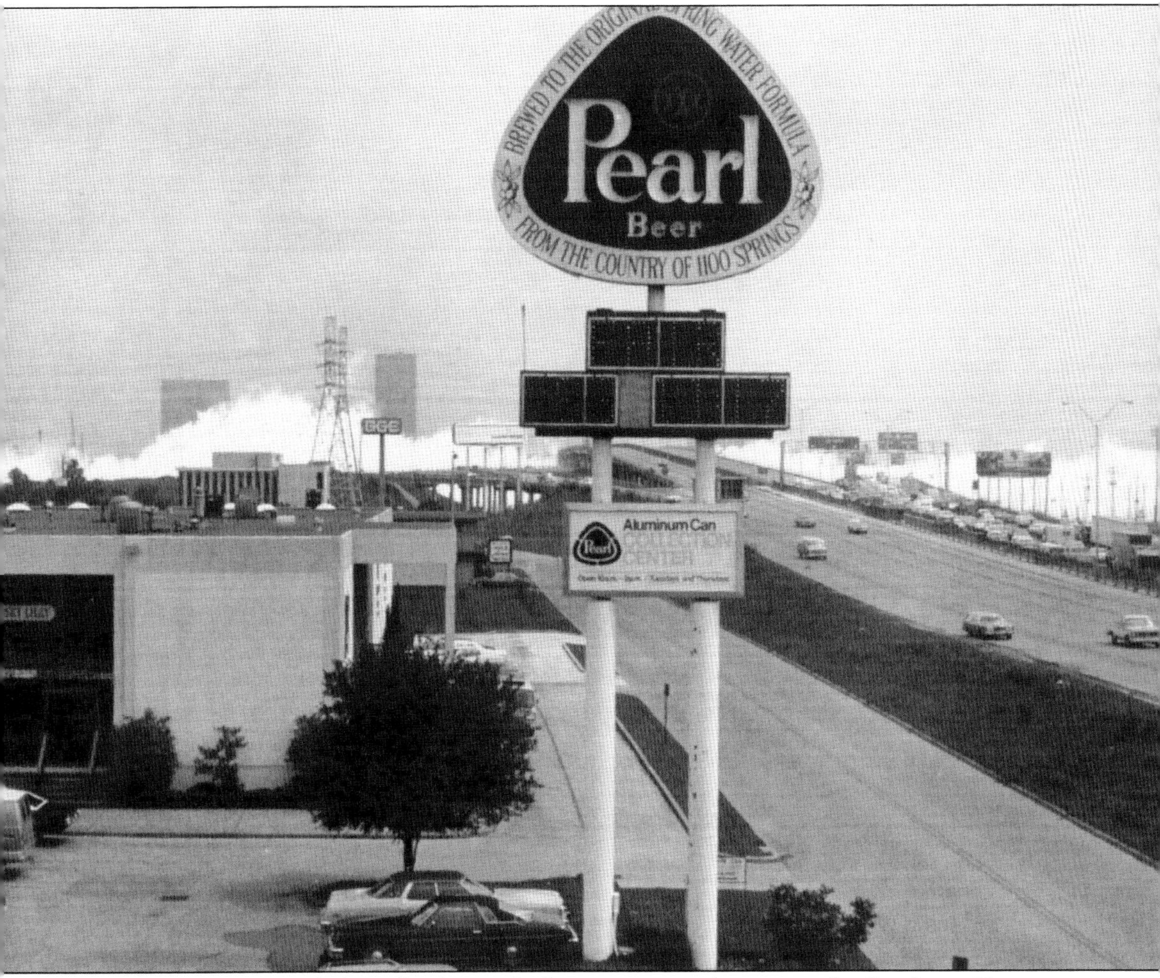

On May 11, 1976, a truck driver hauling anhydrous ammonia drove onto the Southwest Freeway at a high rate of speed. The tanker crashed through the guardrail, fell from the bridge, and exploded, releasing a cloud of toxic gas into the busy interchange. Along with eight fatalities, more than 150 people were injured, including several firefighters. Several nearby subdivisions had to be evacuated.

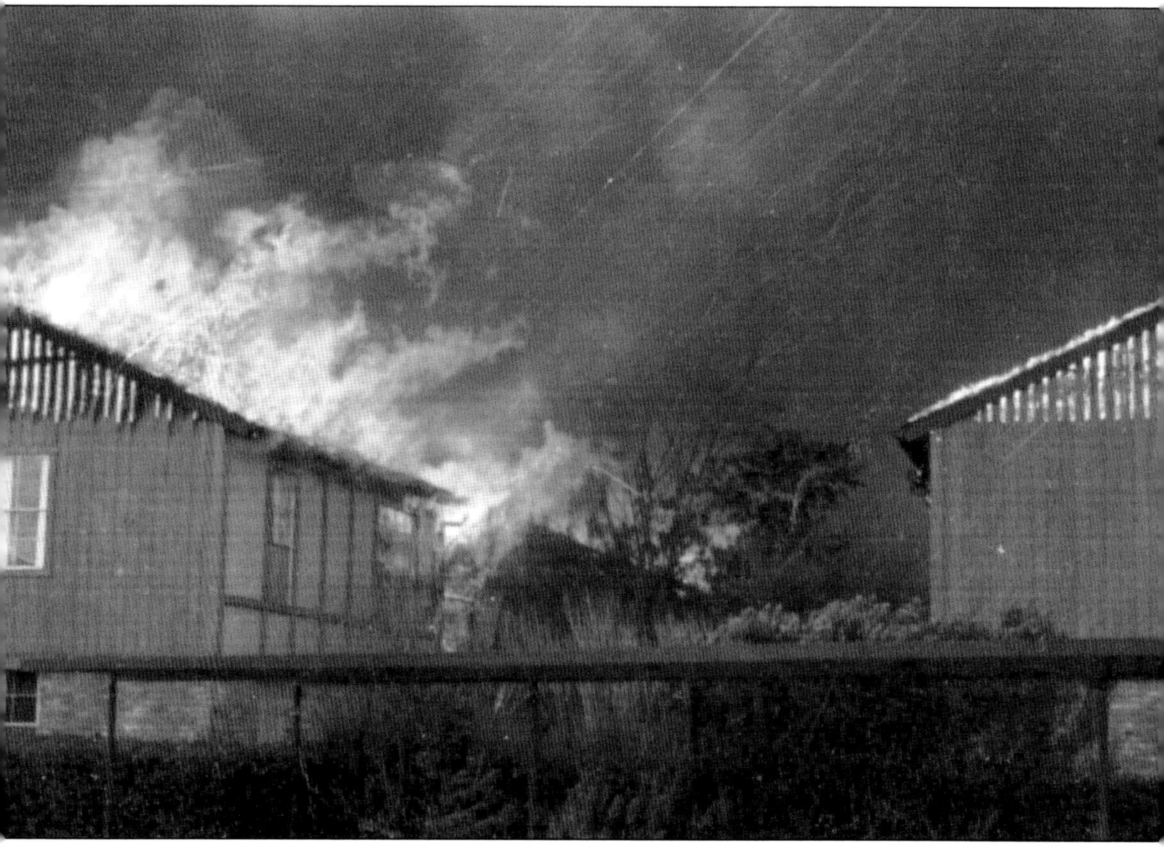

Houston's largest fire since 1912 occurred at the 55-acre Woodway Square apartment complex on Houston's west side at 1501 Winrock Boulevard. On Tuesday, July 31, 1979, a small fire began on one of the building's roofs in the upscale Briargrove area near the Galleria. There had been five multiple-alarm fires in the complex since 1971, but all had been contained at the building of origin.

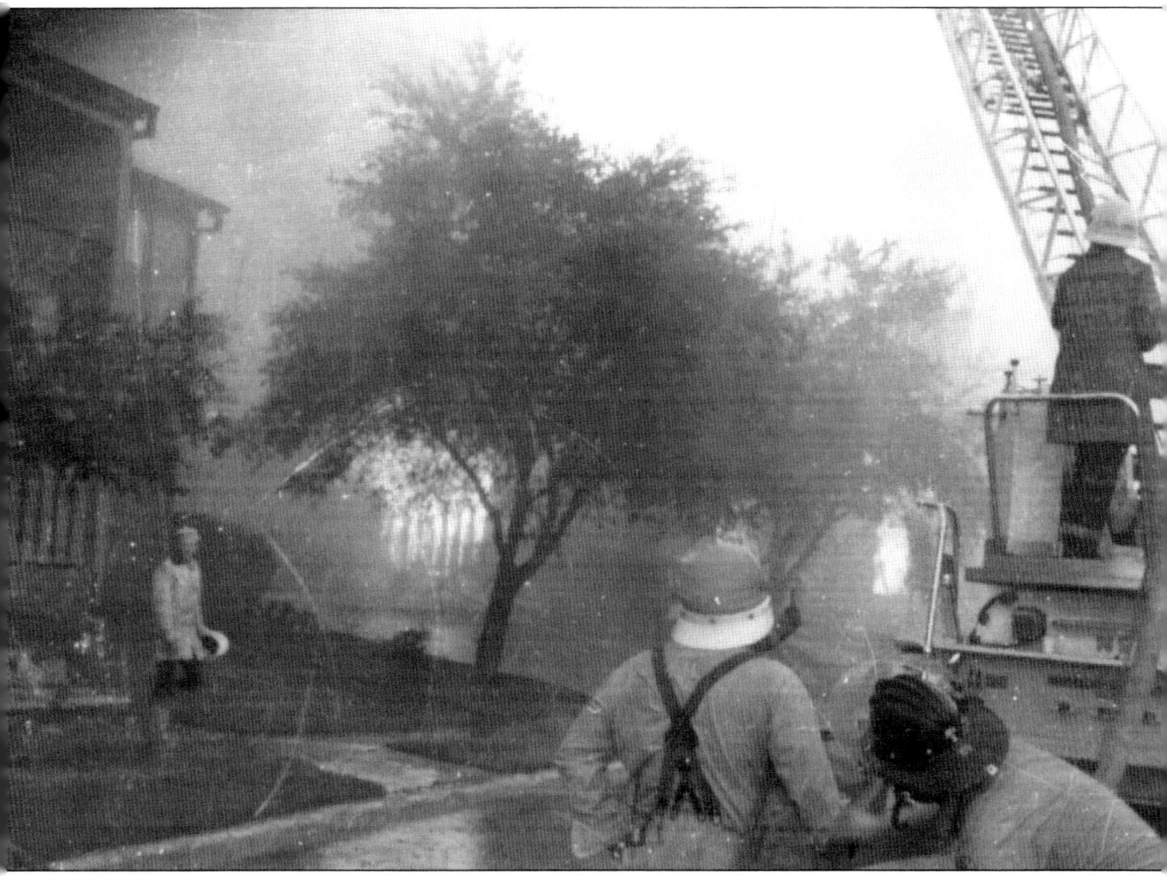

A steady and dry southerly wind, along with the wood-shingle roofs, spread the Woodway Square fire quickly. Dispatchers sent box-four engines, two trucks, and a district chief when the fire was reported that afternoon. It took nearly four hours to control the blaze. In an area equivalent to 10 city blocks, 324 apartments were destroyed and over 100 others were damaged. Thankfully, no one was killed or seriously injured.

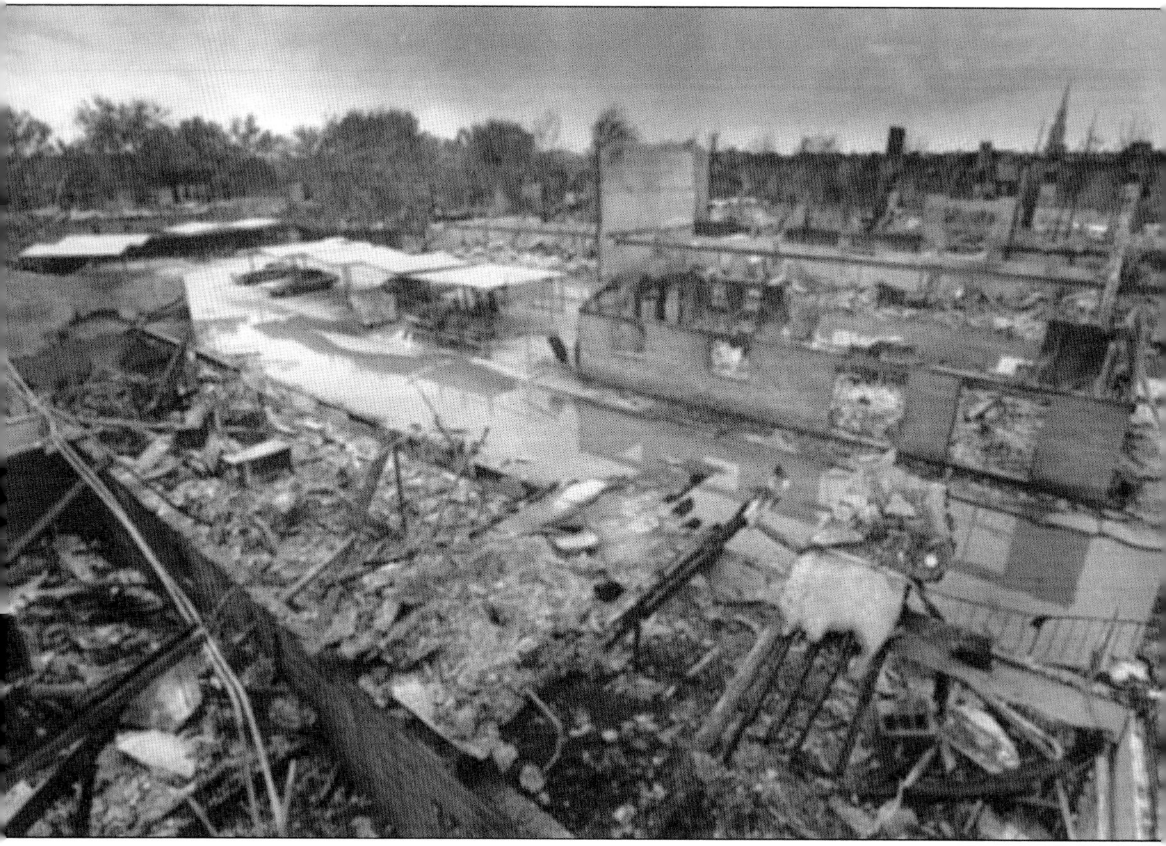

A $30 million property loss was estimated as a result of the Woodway Square blaze. The federal government declared Woodway Square a disaster area and gave victims low-cost loans to cover uninsured losses. The Houston City Council, which had been debating an ordinance against using wood-shingle roofs in multiple buildings, passed the ordinance the next day. Several other building-code amendments were passed later in 1979 that further helped stop fire spread in multiple dwellings.

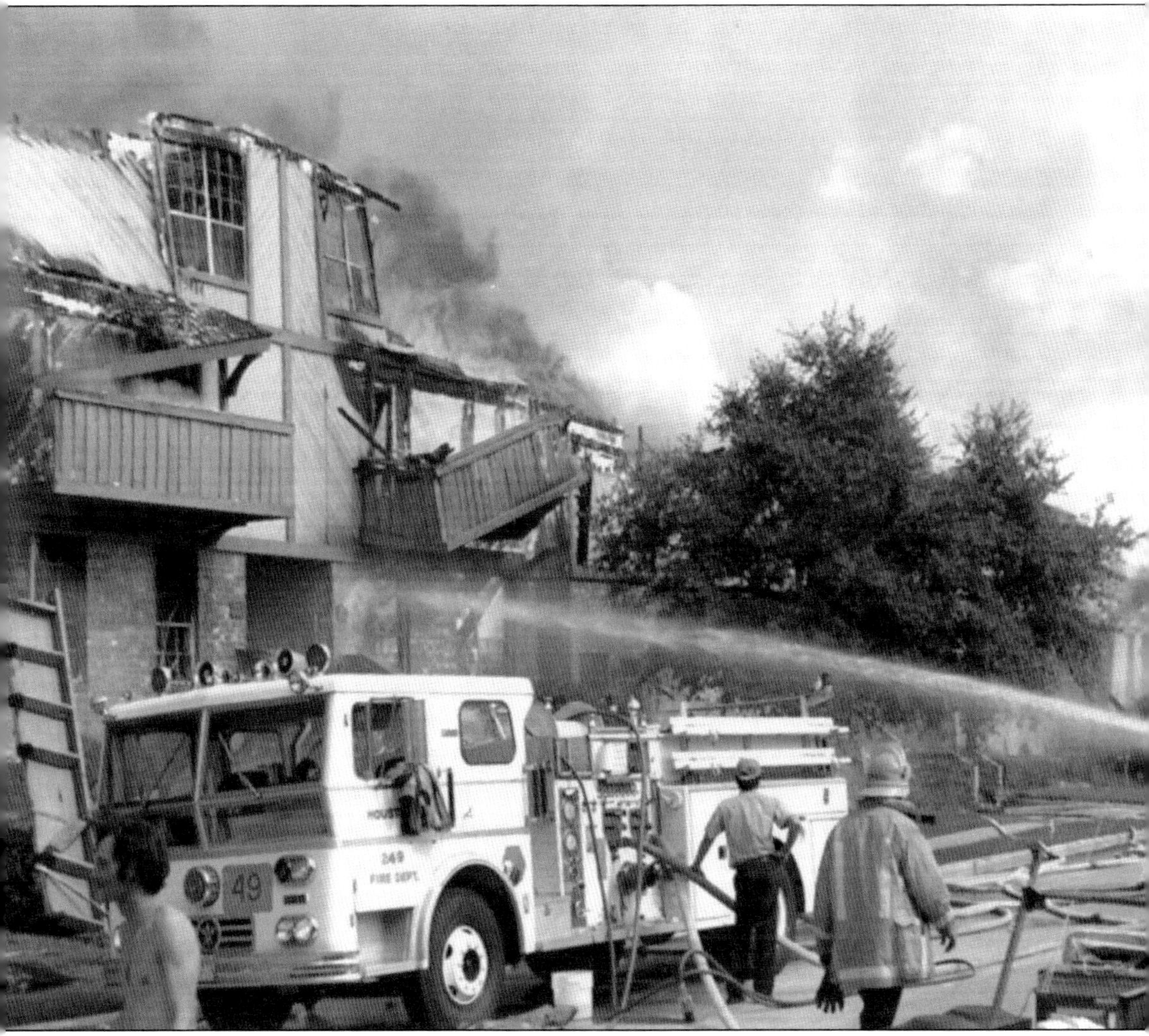

At that point, it was the only time a seven-alarm call had been made. The Woodway Square buildings that remained standing in 1979 were town down in the early 1990s, and new buildings were erected. The new structures were made with composition roofs; neither wood shingles nor shakes were used. The Woodway Square name continues to be used by this different apartment complex on the same site.

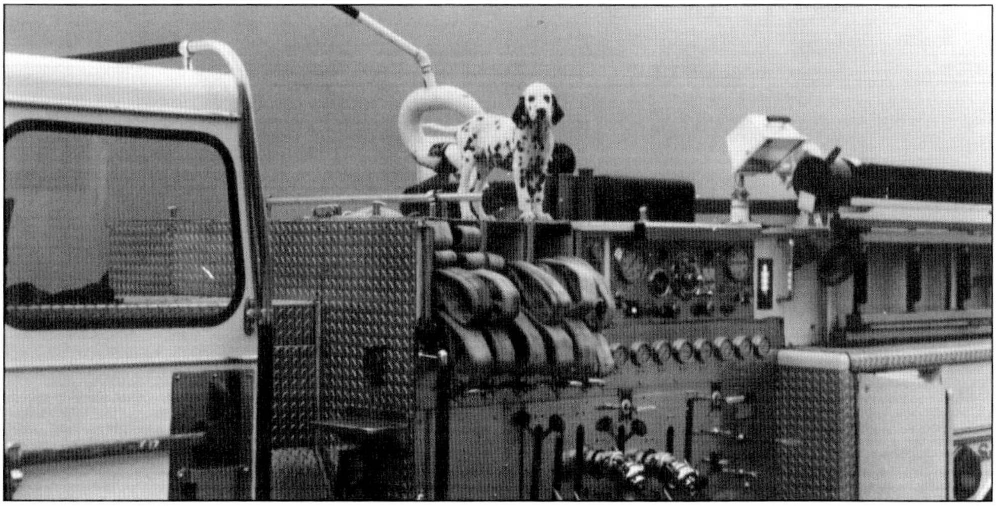

Today, the Dalmatian serves as the firefighter's mascot. In the days of horse-drawn apparatus, they provided a valuable service, due to their affinity for horses. The Dalmatian would run alongside the horses or beneath the cart axles to help keep the horses calm. The Dalmatian also acted as a guard dog. Horses were replaced as vehicles became motorized, but the Dalmatian remains.

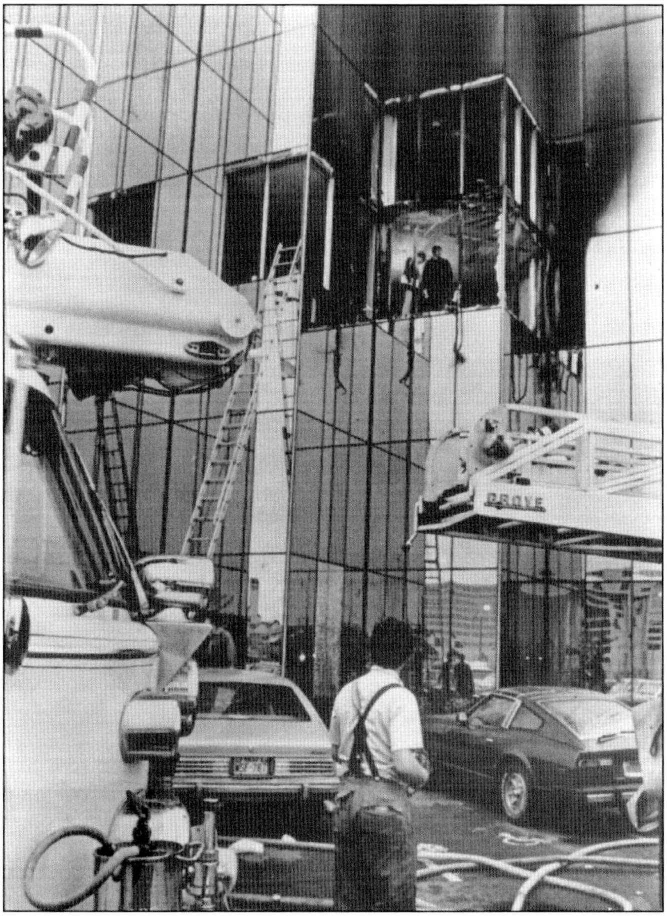

In March 1982, a small fire broke out in the two-year-old Westchase Hilton Hotel. In the blaze, 12 people died, reportedly due to the inadequacy of escape routes and because staff had reset the alarm system, preventing transmission of information to the fire department. As smoke filled the corridors, people found it hard to reach the stairways at the end of the halls and escape to safety, thus dying from suffocation.

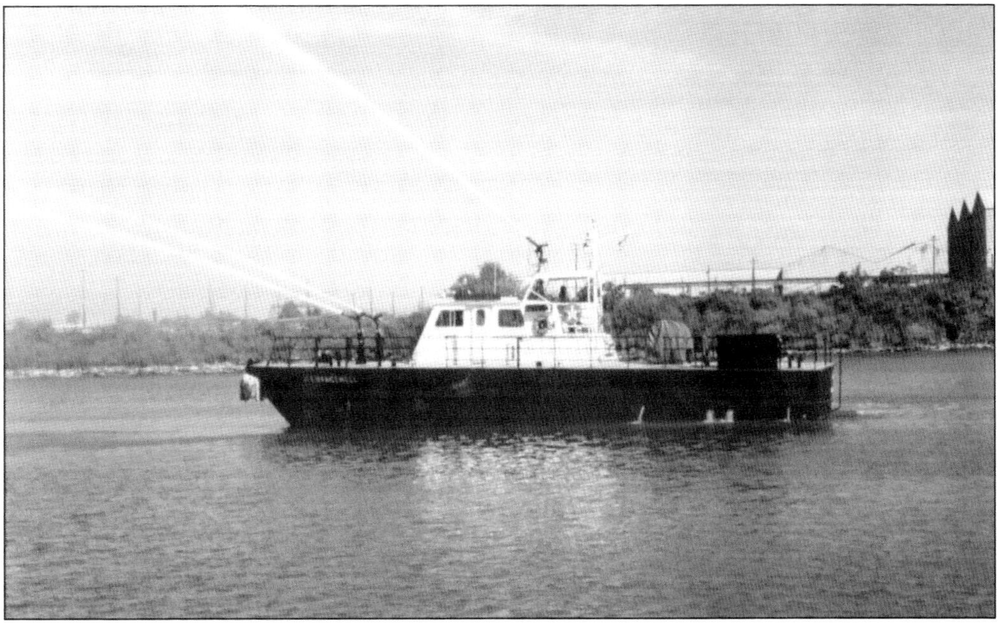

In 1983, two new fireboats, the *J.S. Bracewell* and the *H.T. Tellepsen*, joined the ranks. The aluminum vessels, built by Swiftships of Morgan City, Louisiana, cost $857,000. They were smaller and faster than the three previous fireboats. The boats replaced the *Captain Crotty* in the spring of 1983, one going to the Manchester dock, and the other operating at Barbours Cut near LaPorte.

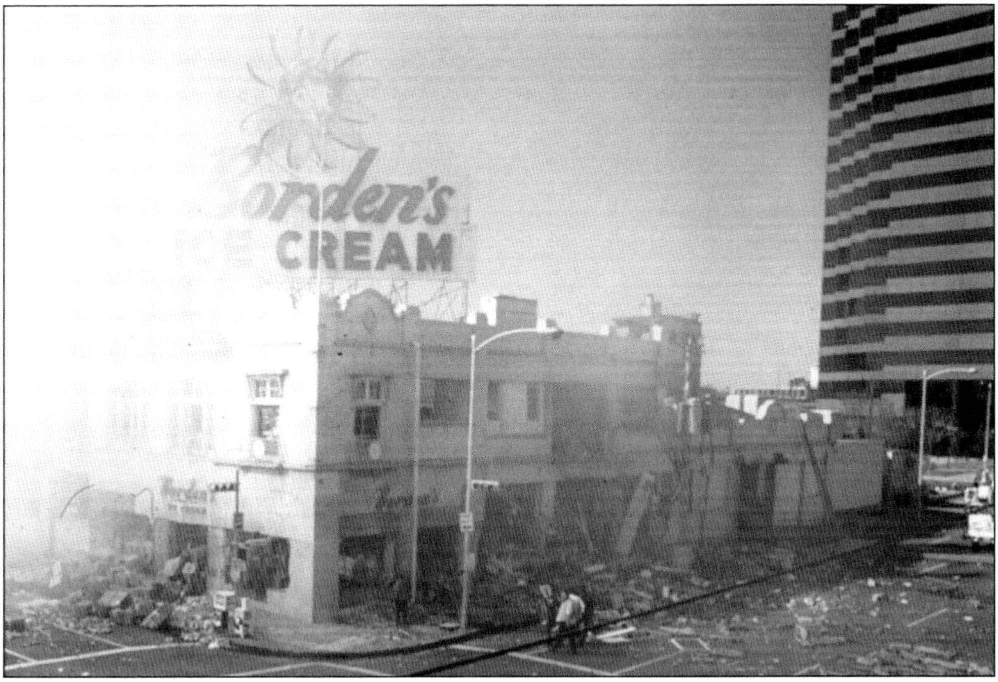

Firefighters were called to the scene of an anhydrous ammonia leak at Borden's Ice Cream plant at Milam and Calhoun Streets early in the morning of Sunday, December 11, 1983. They were helping an employee adjust to an airpack so he could take them to a shutoff for the leaking valve when the building exploded. No one was seriously injured. The site is now Houston Fire Station 8.

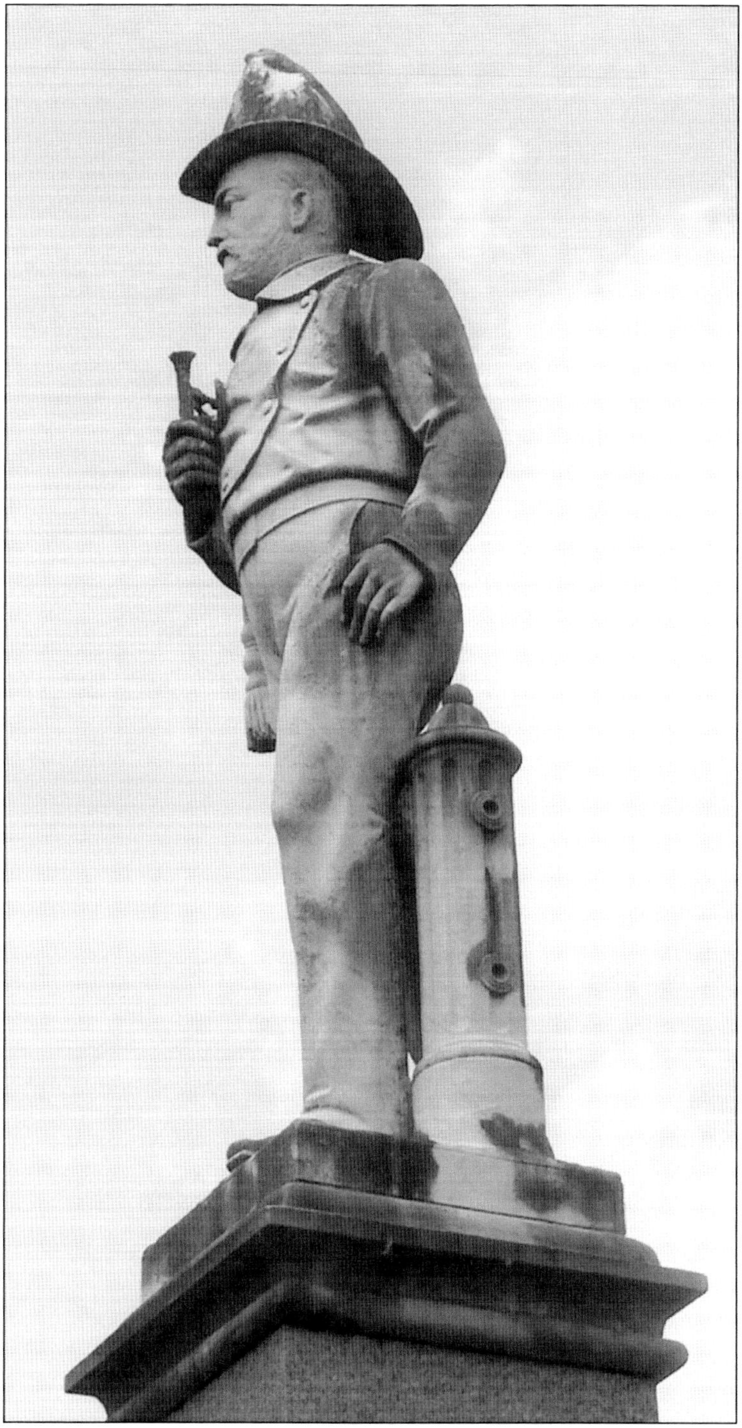

Houston's volunteer firefighters bought a plot in Glenwood Cemetery in 1888 and soon commissioned a monument, which included a base and a statue of Robert Brewster, the oldest living volunteer fireman in Houston. In 1976, city officials moved the statue to fire department headquarters. A move back to Glenwood Cemetery happened in 1992, with a rededication that October.

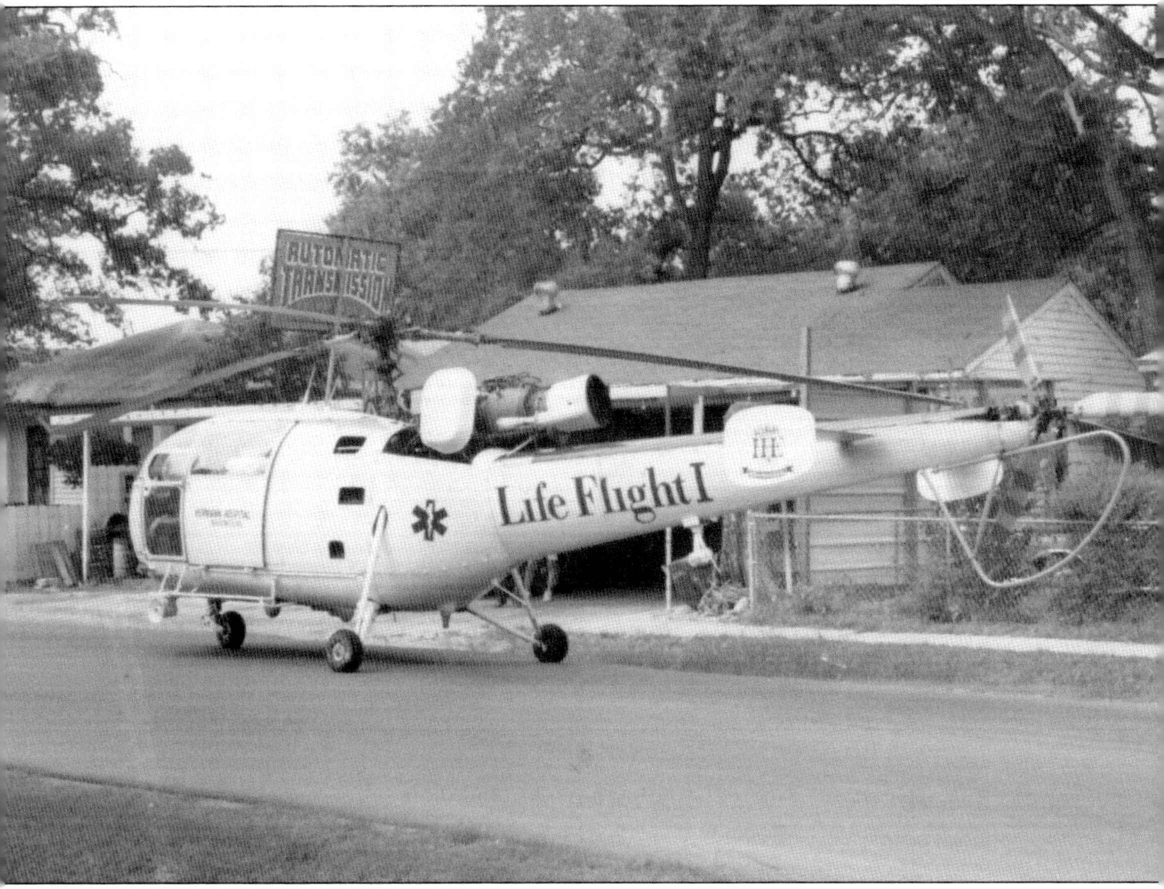

While not under the auspices of the fire department, Hermann Hospital's Life Flight has been an advantageous component for first responders. It was Texas's first civilian emergency air ambulance service. It was first used in the early 1990s at the Phillips 66 refinery plant explosion in Pasadena. These helicopters, placed strategically throughout Houston, average 300 flights per month.

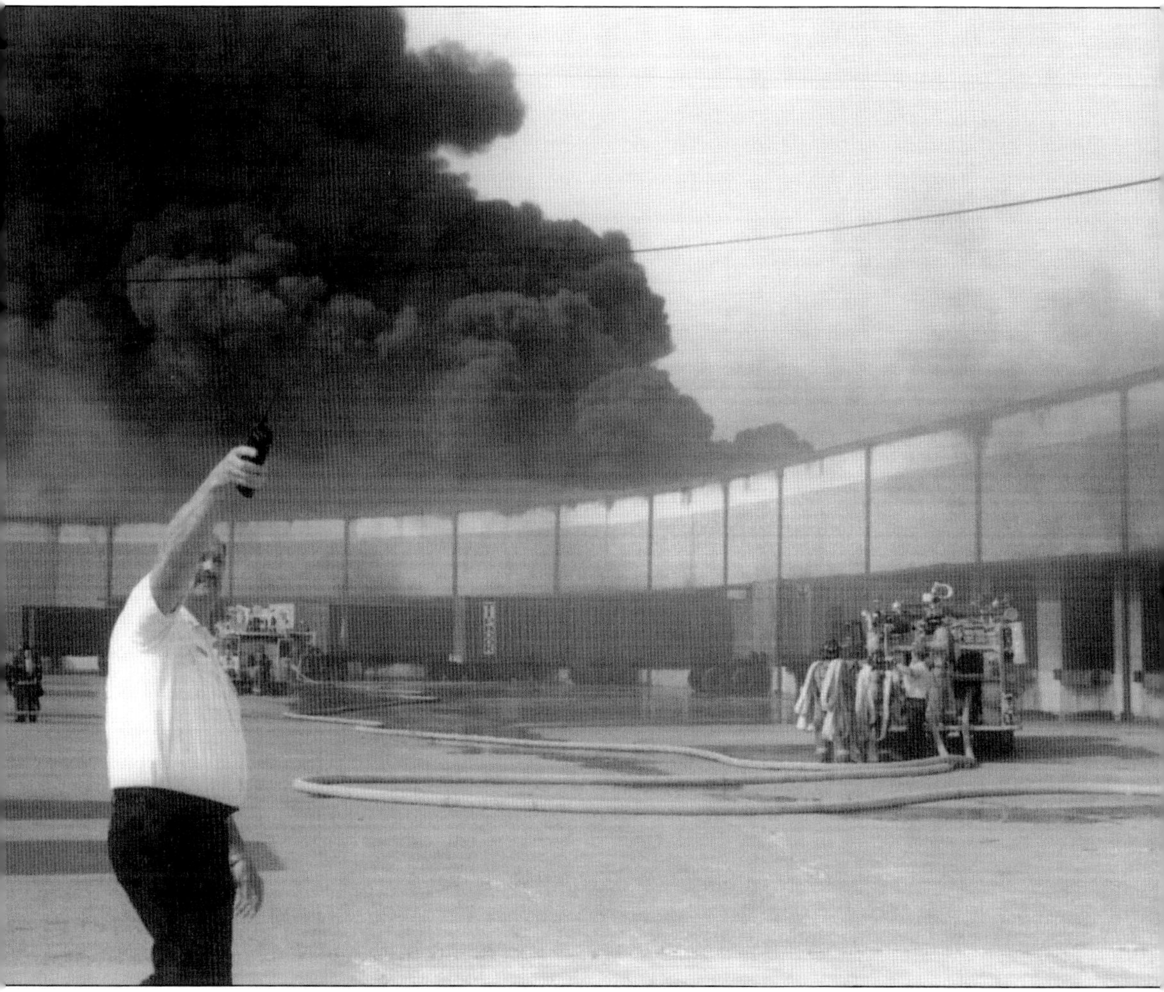

On June 24, 1995, a fire at the Houston Distribution warehouse at 8550 Market Street developed into the first seven-alarm fire since the 1979 Woodway Square blaze. Firefighters were deployed for an interior attack, but they had to retreat when burning chemicals stored inside poured into the parking lot surrounding the complex. Here, District 19-B chief Phil Hudgens directs third-alarm companies into the warehouse's truck parking area.

The Houston Distribution warehouse fire was extremely dangerous, due to stored plastics, poisons, and flammable liquids. It was one of the few times an apparatus was destroyed. Ladder 19 became a victim when hundreds of gallons of burning liquid flowed down a dock and surrounded it. The fire took nearly 18 hours to put out.

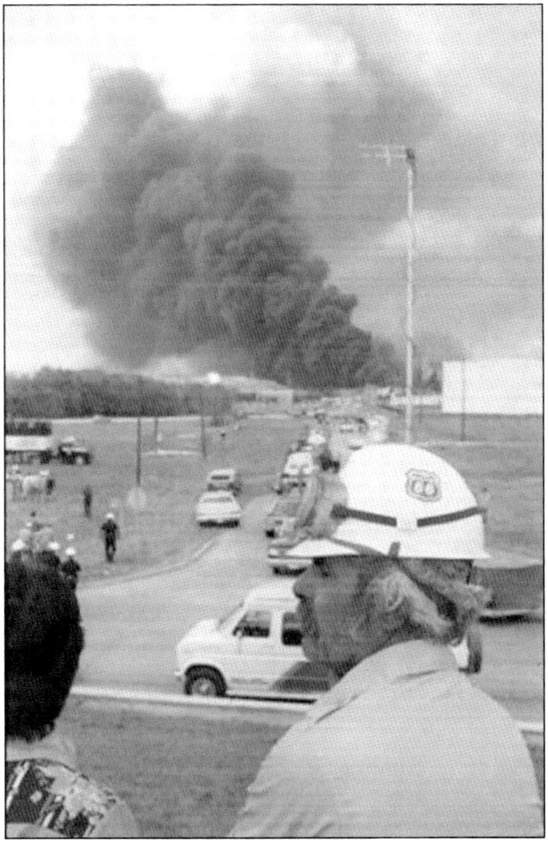

Less than two weeks after the Houston Distribution warehouse fire, on July 9, firefighters returned to see the section they had struggled to save nearly fully engulfed. The remaining portion of the warehouse was completely destroyed by flames in what, itself, became a four-alarm fire. In 1995 and 1996, there were 13 fires that required four alarms or more, the highest since the late 1970s.

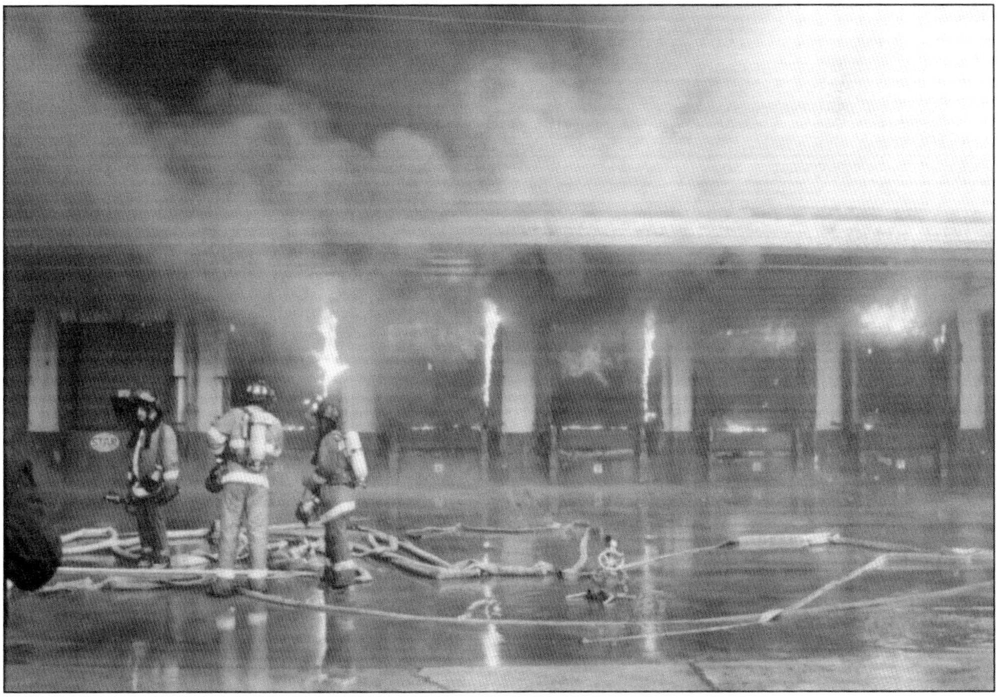

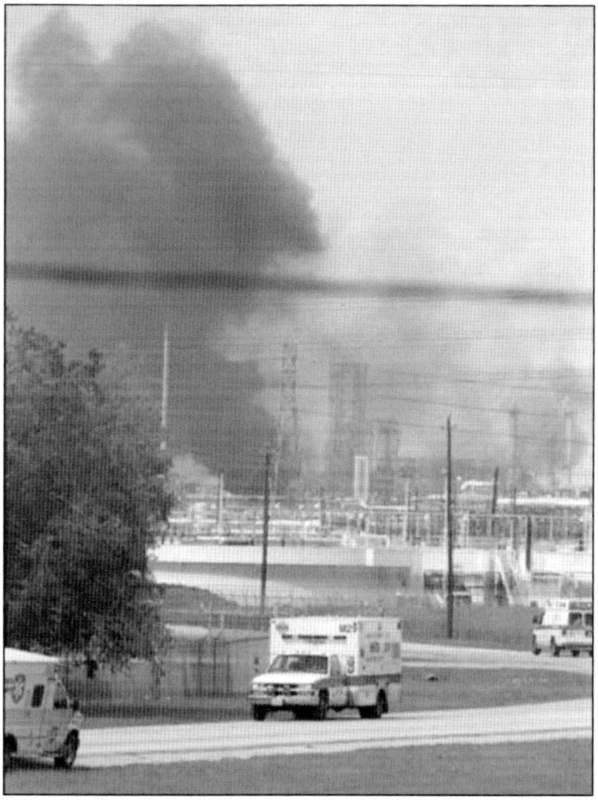

Houston firefighters responded on the afternoon of March 27, 2000, to an explosion and fire at the Phillips Petroleum plastics manufacturing and storage plant in Pasadena. The incident was responsible for one death and 71 injuries. The fire produced huge plumes of black smoke that spread over the heavily industrialized Houston Ship Channel and neighboring residential areas.

Washington 8 was organized in August 1894 at 1307 Crawford Street. The building deteriorated badly over the next 64 years, and was demolished in 1958. The arch over the apparatus door had cracked, and some of the men feared it might collapse. Then, two newer stations were erected on the site until 2001, when the final station at that address was demolished to make way for an arena.

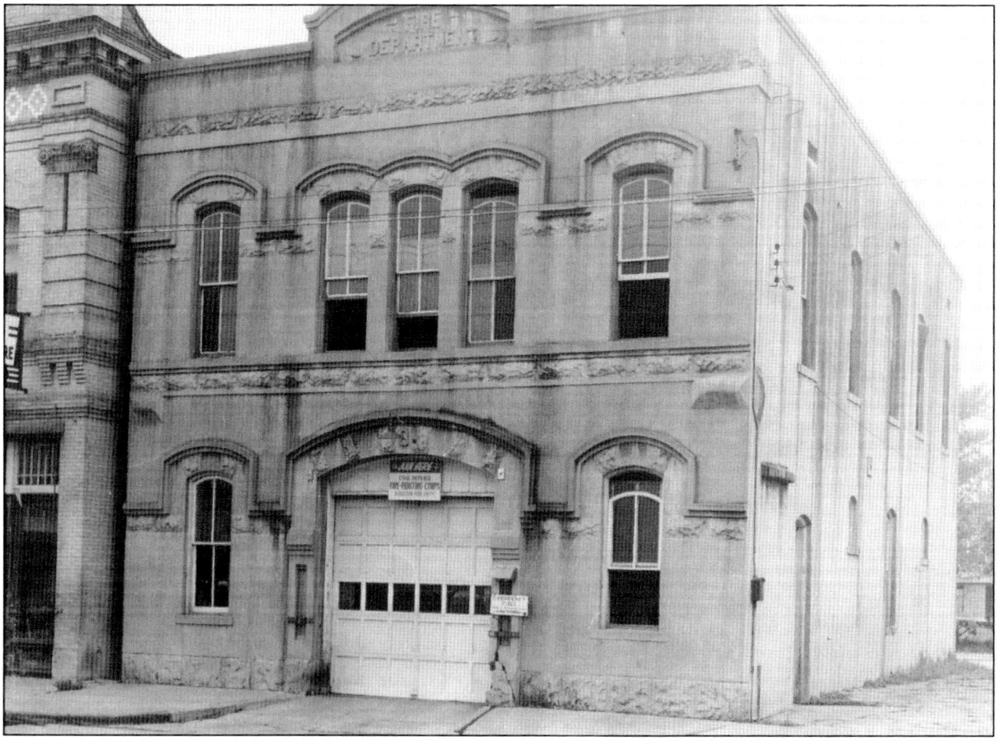

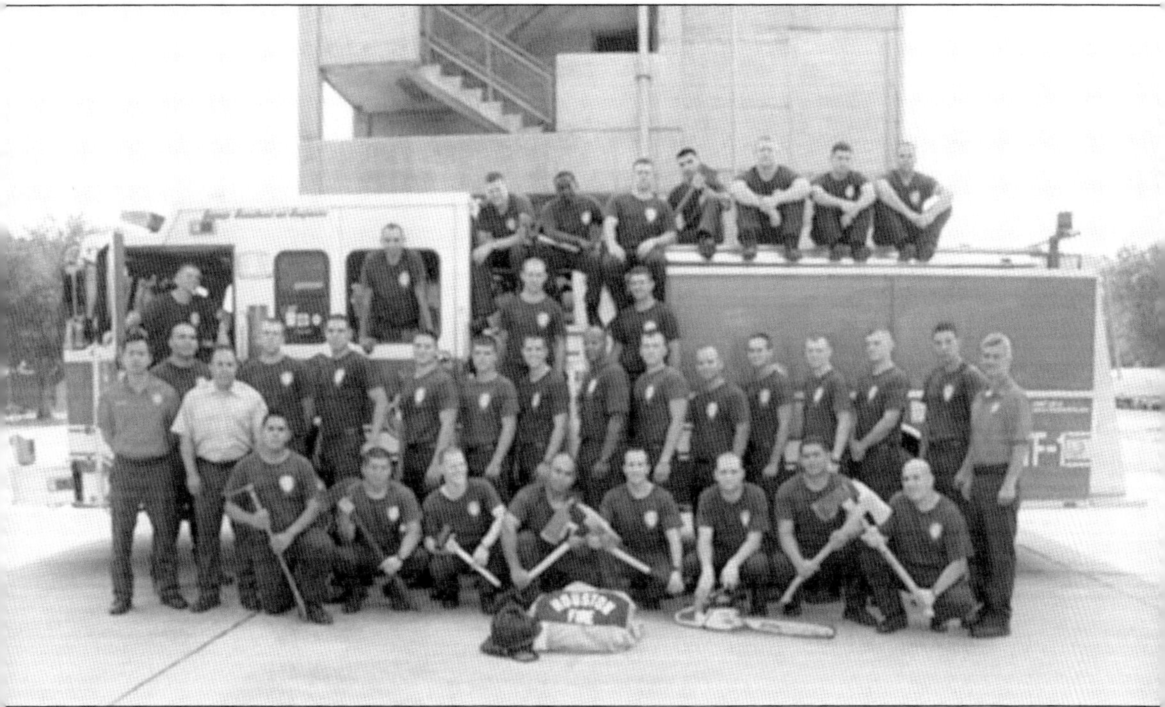

Following the events of September 11, 2001, and Hurricanes Katrina and Rita, HFD recognized the need to become a regional training center. Looking to fulfill the needs of departments beyond Houston, it renovated the Val Jahnke Training Facility in 2006. Renovations and remodeling include the burn building and drill tower, both of which have been used extensively since 1969.

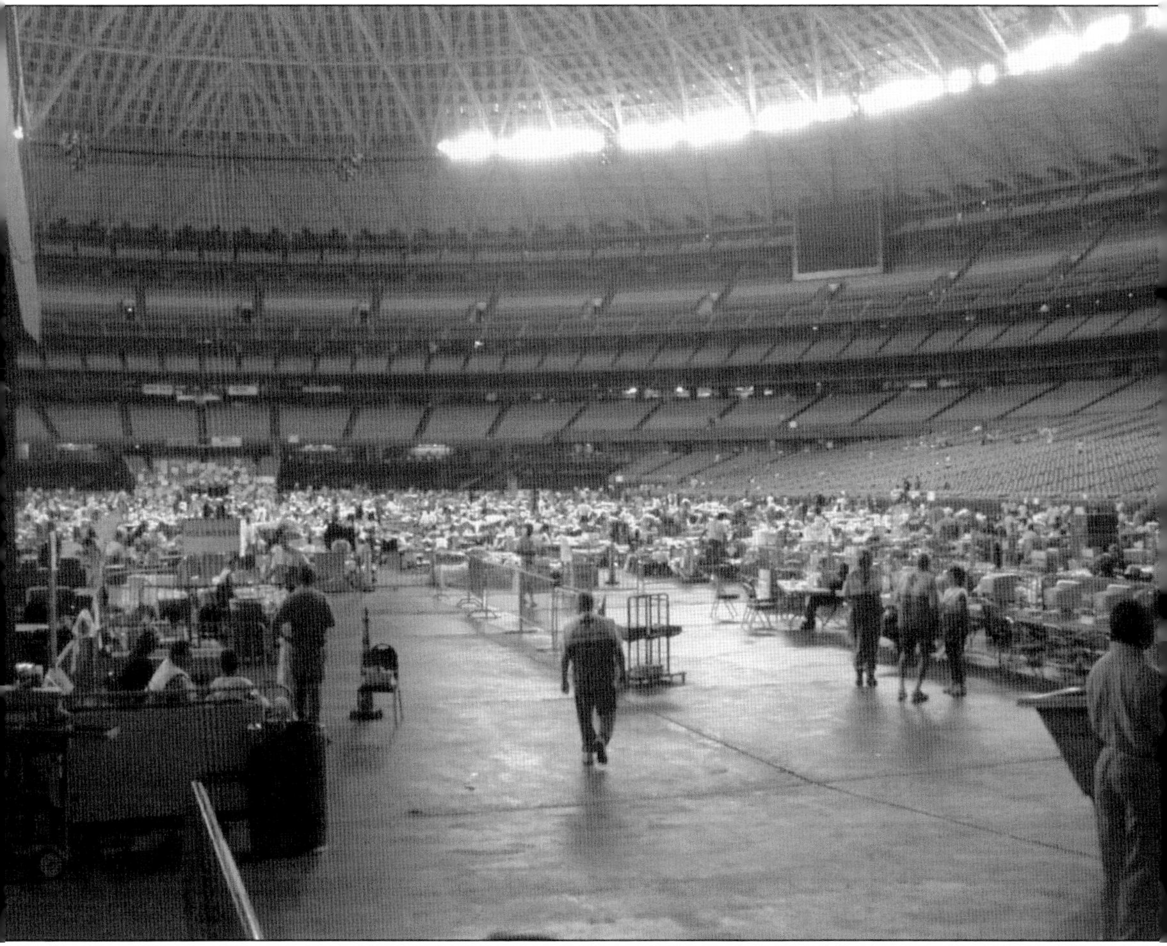

After Hurricane Katrina hit the Gulf Coast on August 29, 2005, nearly 23,000 evacuees, many ill or injured, were transported to Houston. The city's regional disaster plan was quickly implemented, and, within 19 hours, the Astrodome/Reliant Center Complex mega-shelter was created. A medical facility was constructed within the Reliant Center. The HFD was a critical component in this operation.

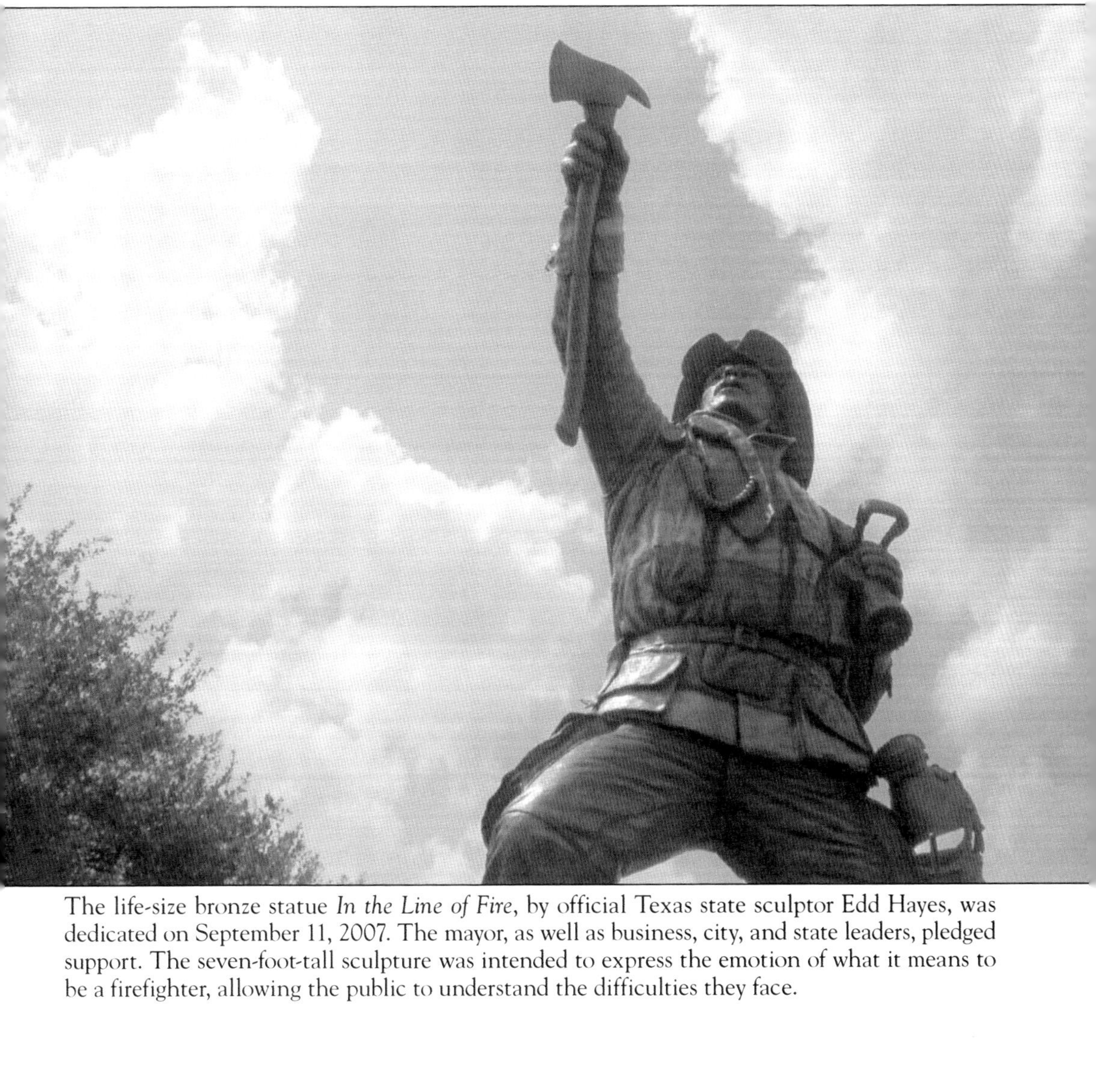

The life-size bronze statue *In the Line of Fire*, by official Texas state sculptor Edd Hayes, was dedicated on September 11, 2007. The mayor, as well as business, city, and state leaders, pledged support. The seven-foot-tall sculpture was intended to express the emotion of what it means to be a firefighter, allowing the public to understand the difficulties they face.

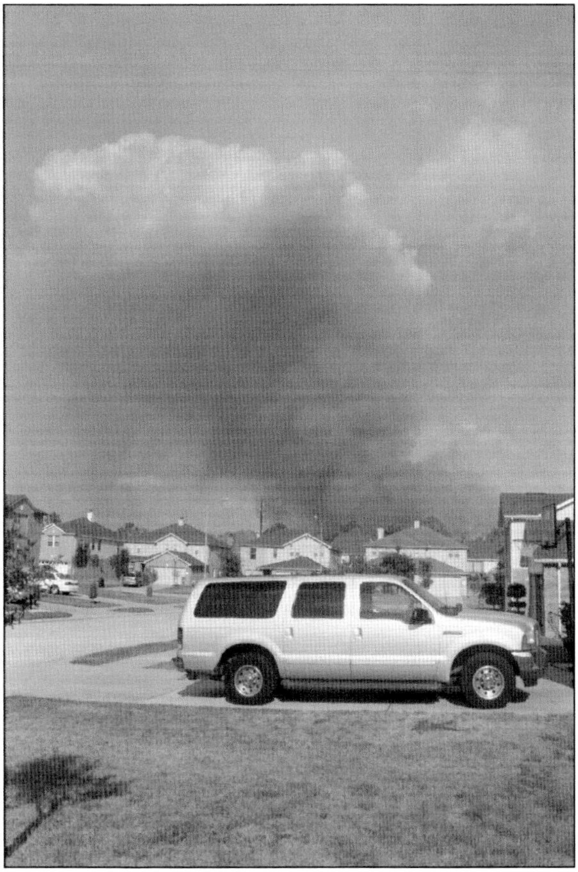

Extreme drought plagued the southwestern United States in 2011, including Houston. Firefighters were dispatched to a wildfire that began in George Bush Park, near Westheimer Parkway and Highway 6, on September 13. The blaze quickly grew, and the firefighters moved to the north levee near Interstate 10 to wait for the fire to come to them in order to stop it. In the end, it consumed 1,500 acres.

Firefighters answered a call at the Southwest Inn, located in the Greater Sharpstown district along the Southwest Freeway, on May 31, 2013. The blaze began in the Bhojan Indian restaurant during lunch. The ensuing fire claimed the biggest casualty loss for the HFD since its inception. The five-alarm fire resulted in the deaths of four firefighters. In addition, 13 others were injured battling the blaze.

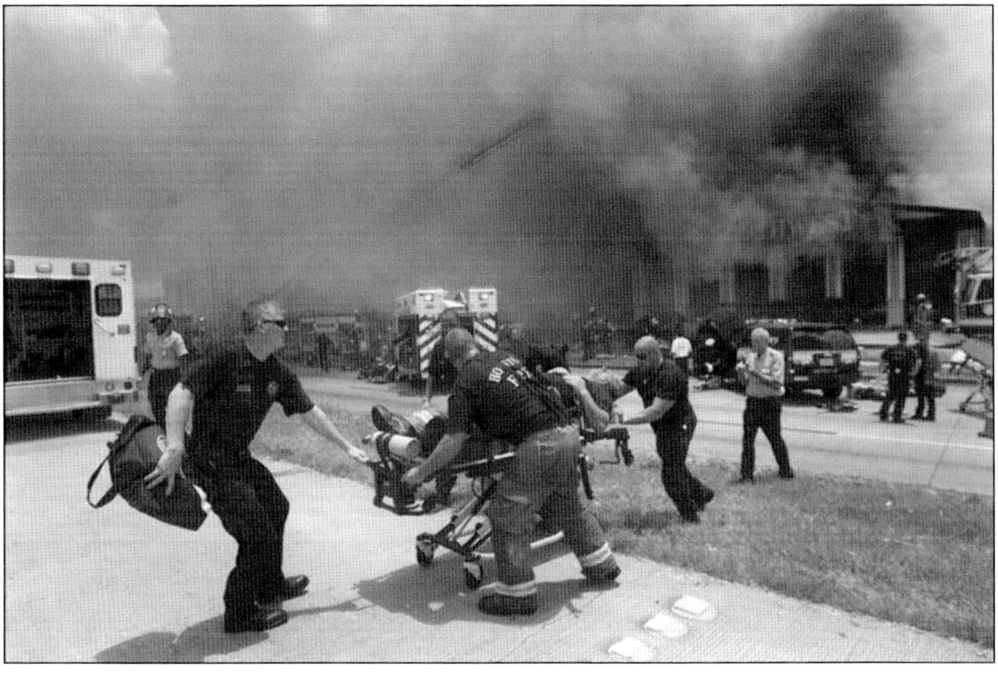

Five

HOUSTON'S FIRE CHIEFS

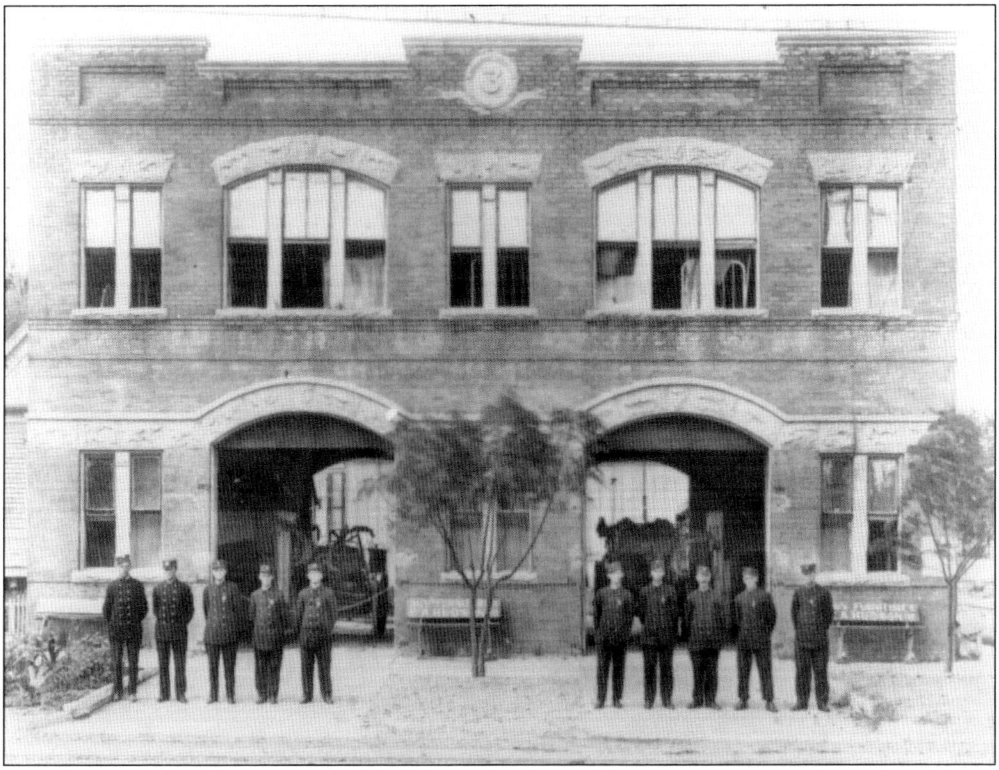

The Houston Fire Department has grown exponentially since coming under the guidance and organization of the city of Houston. As the city has grown and changed, so, too, have the demographics, the landscape, the technology, building materials, approaches to fires and other disaster events, and the approach to the fire service in general. Due to the changing complexities in Houston, the skills and expertise needed by those who lead the department have evolved. Station 3, seen here, also known as "Stonewall," was one of Houston's original fire stations. Built in 1905, it eventually outlived its usefulness to the fire department and has most recently served the city as an event venue.

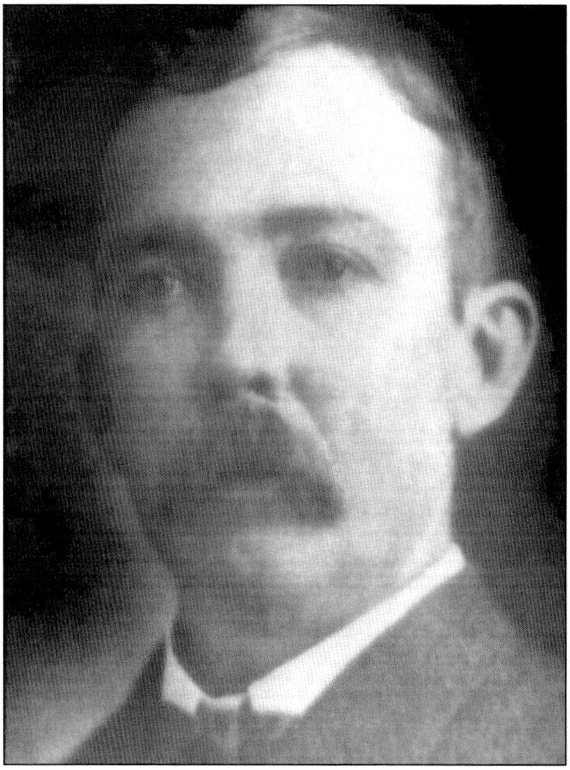

Thomas Ravell was appointed the first fire chief of the fully paid department in 1895 after having been elected chief of the volunteer department in 1894. Under Ravell, the department established the first rules of conduct, scheduling, and staffing. Ravell oversaw 44 firefighters, seven of the volunteer fire stations, and the department's apparatus, nearly all of which had been under the ownership of the volunteer stations.

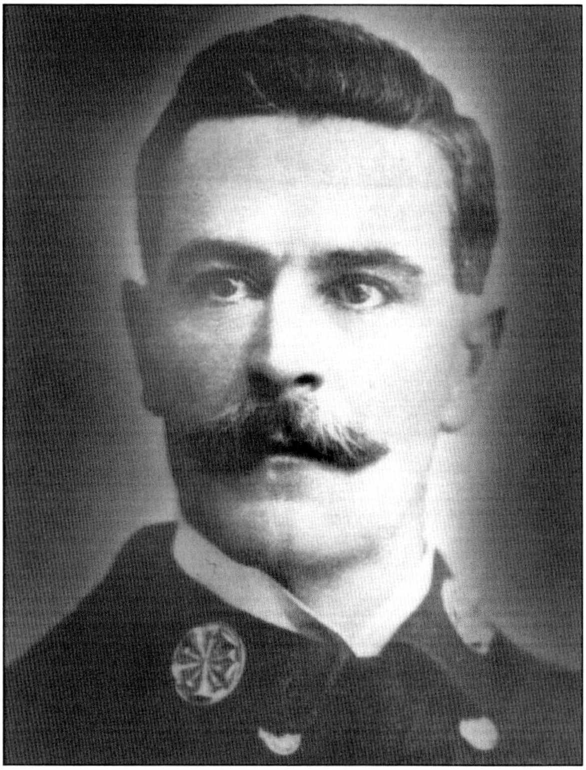

James Hussey had served Houston as a volunteer firefighter. The department he took over from Ravell had grown to 61 firefighters in nine stations. During Hussey's leadership, the Houston Relief Association of the Paid Fire Department of the City of Houston was organized. Its purpose was to help members who were distressed, injured, sick, or disabled. One representative from each fire company made up the board of directors.

Jack Arto, another former volunteer, was widely admired by the firefighters, who called him "Little Chief" due to his short stature. The year Arto took office, oil burners were installed on the steamers, firefighters organized a union affiliated with the American Federation of Labor, and the city opened a new Central Station. The city also abolished the ward system and established a commission form of government.

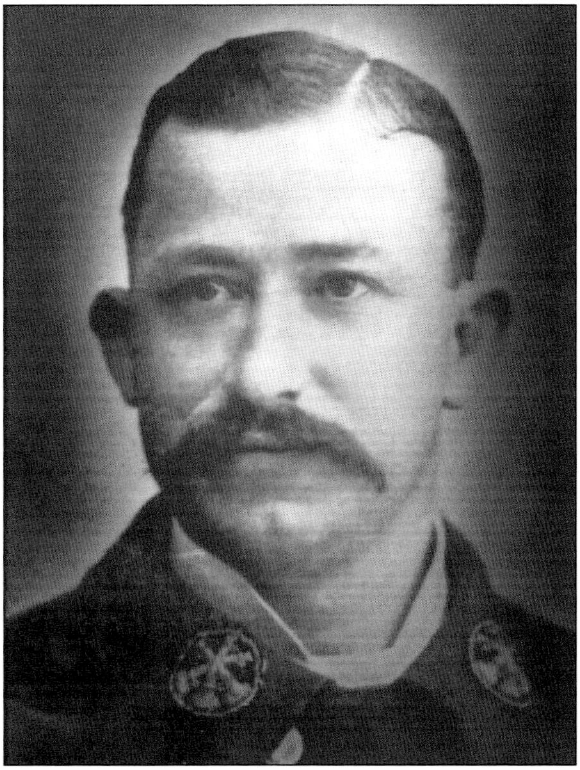

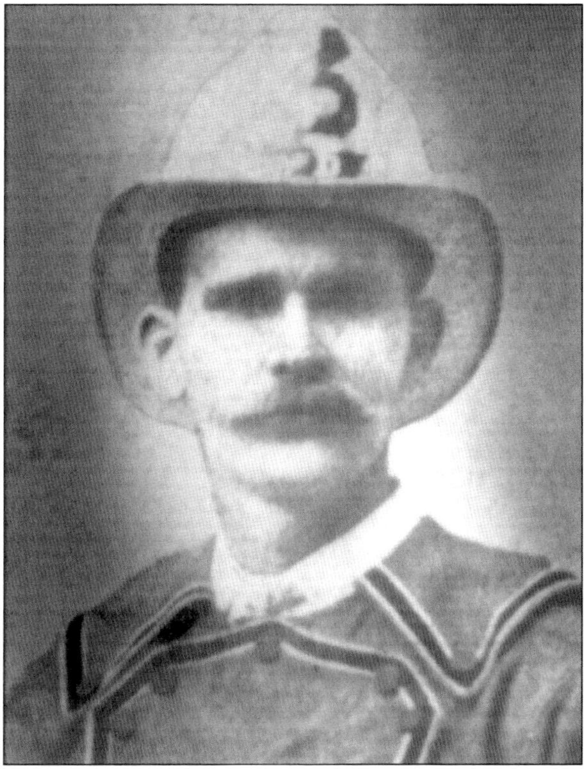

Tom O'Leary, who had served with the volunteer fire department, was named fire chief in 1905 by James Appleby, the first commissioner of fire and police under the new city government. Under O'Leary, Houston purchased a private water system to serve the city, and the union folded in 1906 after rumors of the chief's unfavorable viewpoint on labor unions.

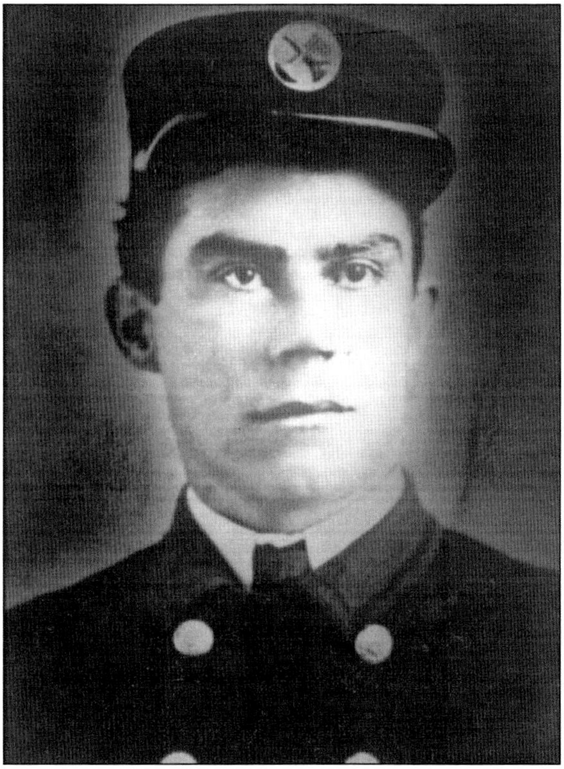

Reginald "Kid" Ollre, a former volunteer, became the fifth fire chief following the death of O'Leary. In 1910, a new position of fire marshal was created, and the fire department purchased its last steamer, bringing the total number of steamers to eight. With over 100 firefighters, the total number of alarms answered by the department in 1911 reached 999.

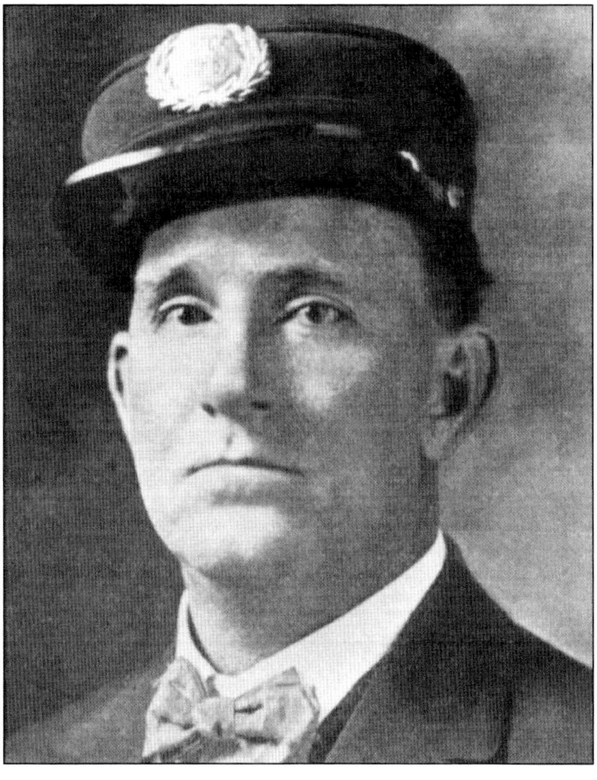

Fred Seibert replaced Chief Ollre late in 1912. Seibert freed firefighters from nighttime outside watch duty. Under his lead, firefighters came under city civil service, and the motorization of fire apparatus began. Seibert was replaced in 1919, but he returned 10 years later. This time, he saw the department issue fire helmets, the organization of the Harris County Emergency Corps, and the formation of Local 341.

William Wells was picked to replace Fire Chief Seibert after voters elected Earl Amerman as mayor in late 1918. During Wells's tenure, a second platoon was created in 1912, with the two alternating monthly between 10-hour day shifts and 14-hour night shifts. A civil-service grading system was implemented in 1921. By that year, all the fire apparatus had been motorized and the horses retired to a city park.

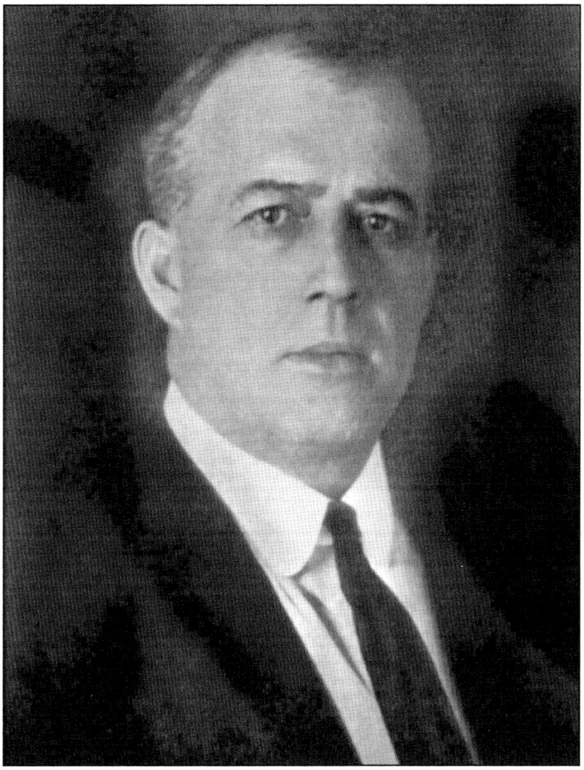

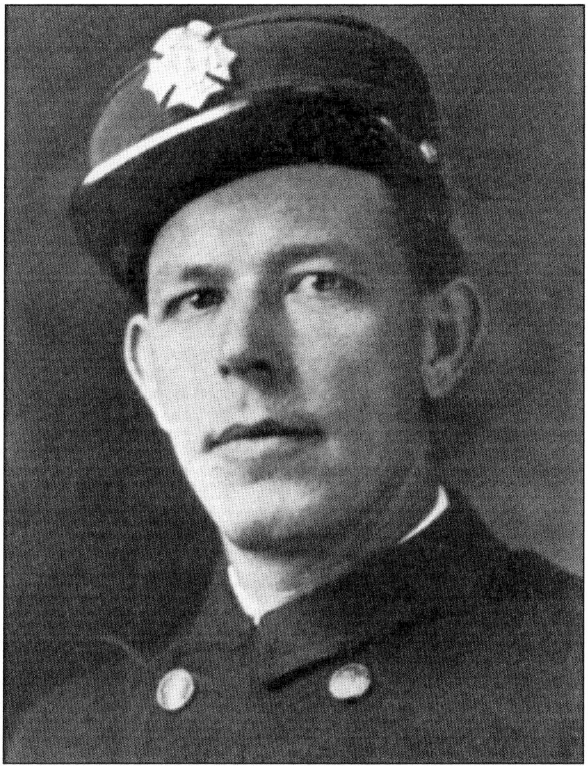

Cornelius "Roxie" Ollre replaced Wells in 1923. Under Ollre, a safety committee was formed to respond to fires and to assist in crowd control and traffic. Firefighters began inspecting potential fire-hazard structures, the first fireboat went into service, and the city was divided into four districts. A training facility equipped with a five-story drill tower and ventilation building was constructed in 1928 adjacent to the Central Station.

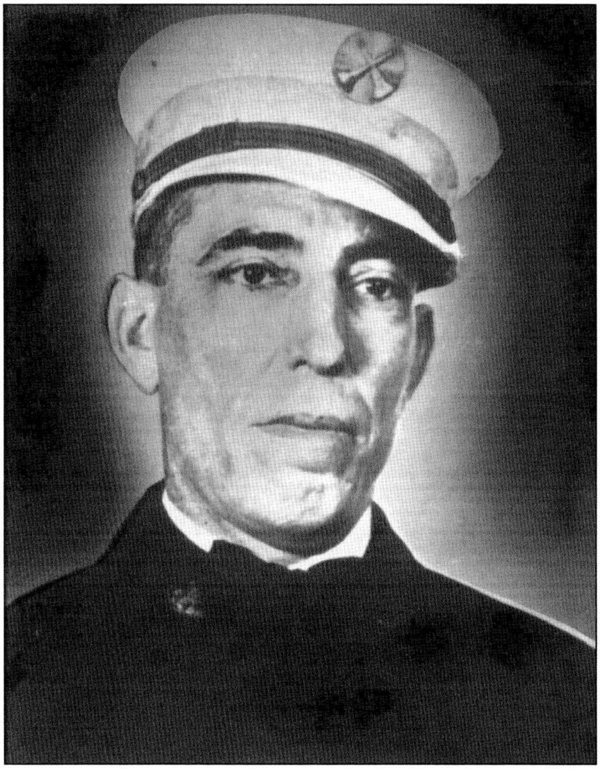

E.B. Hartwell replaced Chief Seibert as fire chief in 1933. Allie Anderson beat H.A. Delery in the city election after Delery served only one term as fire commissioner. Anderson had served 10 years as fire commissioner before he was turned out in the previous election. By this time, the department had grown to 382 men and 26 fire stations.

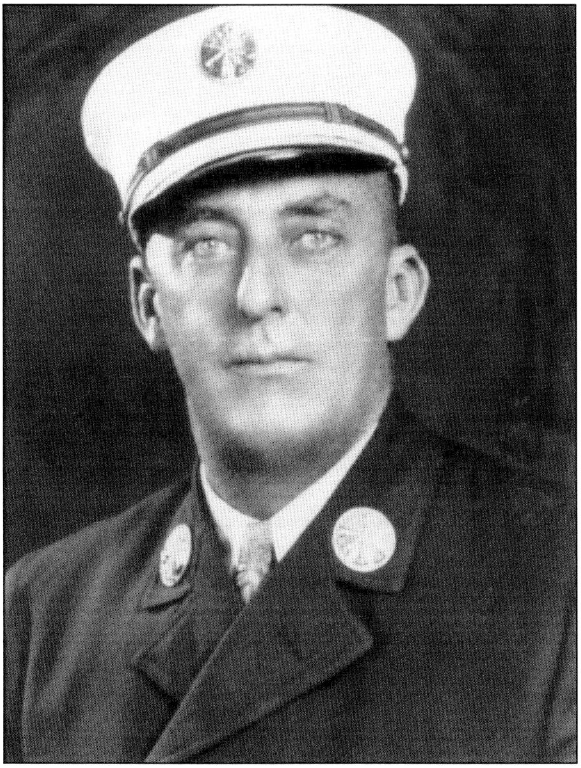

Jasper Davidson was appointed chief in 1935. He put a salvage wagon in service in the late 1930s and ran it out of Fire Station 1. It carried large tarpaulins, called salvage covers, which were used to protect furniture and goods from water during a fire. Previously, salvage work had been performed by the Harris County Emergency Corps. Davidson stepped down in 1939.

Homer Lyles has the longest tenure as Houston's fire chief. During his first stint (1939–1947), firefighters' working time was reduced, the new fire alarm building opened with improved communications systems, and an auxiliary fire force was established. During his second term (1951–1953), Lyles introduced dry-powder fire extinguishers and oversaw the largest purchase of fire apparatus to date. Lyles died of a heart attack in May 1953.

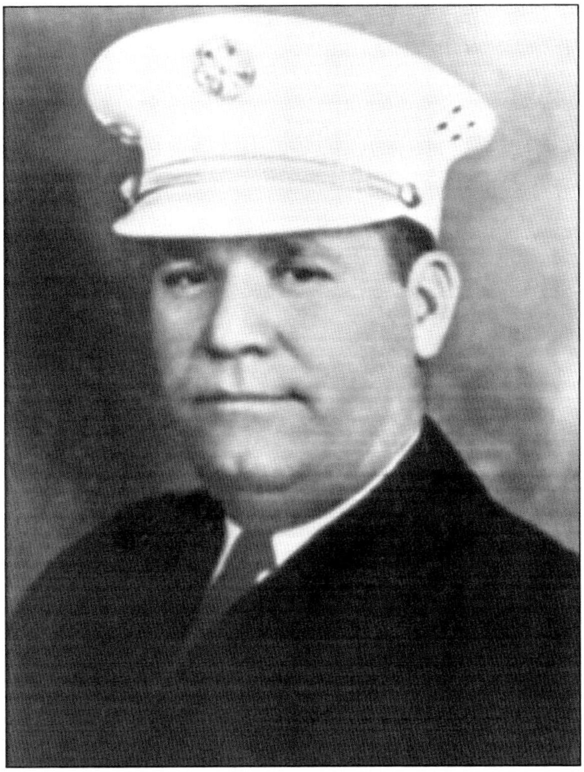

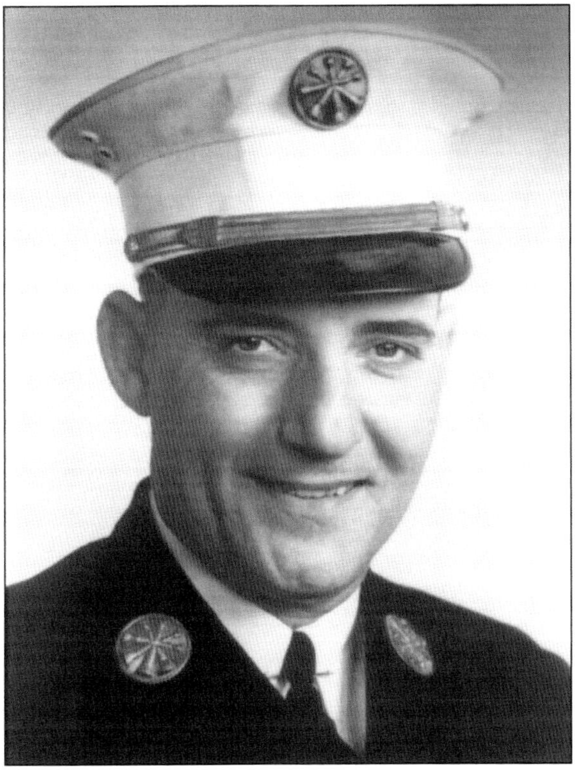

Assistant Fire Chief G.W. Richardson was picked as temporary fire chief after Homer Lyles took a sudden leave of absence in January 1947. The *Port Houston* had recently been placed back in service after public concern following a massive wharf fire, which caused numerous injuries and great damage. By the time Richardson took office, there were over 500 firefighters in service.

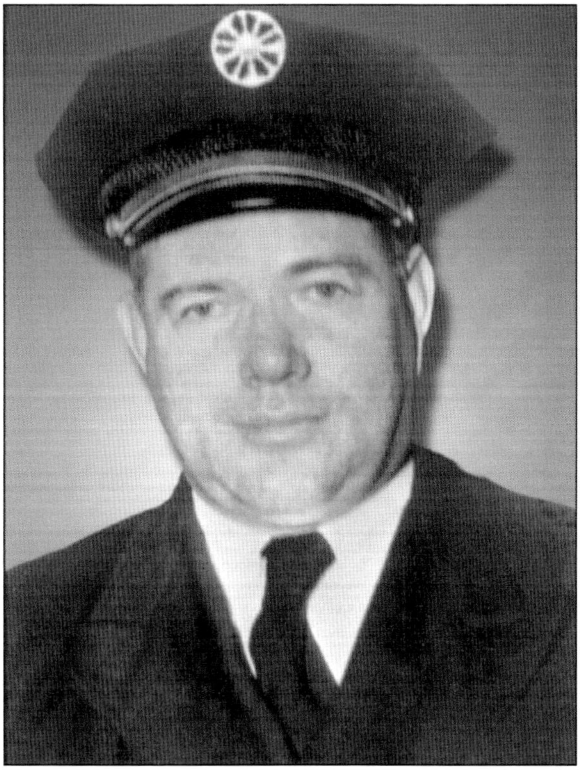

Roy Whittlesey, former fire union president, was named chief in May 1947. At 37 years old, Whittlesey was the youngest fire chief at the time of any major city in the United States. The state legislature approved a bill providing for fire and police state civil service, firefighters began receiving higher ranked pay, and fog nozzles were assigned to fire stations in residential areas.

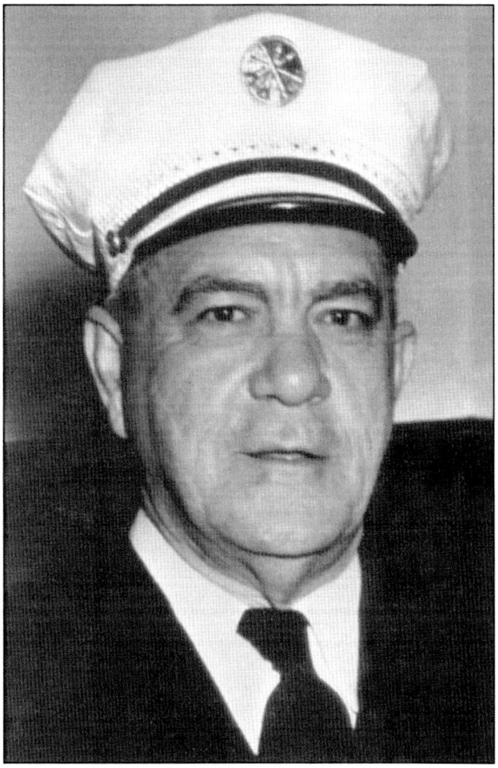

Assistant Chief Charley Middlekauf became chief in 1948 and witnessed a doubling in city size through annexation. Middlekauf implemented reserve fire apparatus and grass-fire trucks into the newly acquired territories. His tenure saw the voters' approval of state civil-service status for firefighters and the Navigation District's absorption of the Port of Houston's fire operations. Middlekauf returned to his former position in 1951, with Thomas Powers stepping in.

Deputy Chief Joseph Lobue was selected chief in November 1953, following Chief Lyle's death. Lobue oversaw the implementation of self-contained breathing apparatus and two-way radios, a huge annexation, the integration of African American firefighters, and the reorganization of the Houston Volunteer Fire Department. Chief Lobue resigned following Mayor Oscar Holcombe's reelection loss and served a short stint again under Louie Welch in 1964.

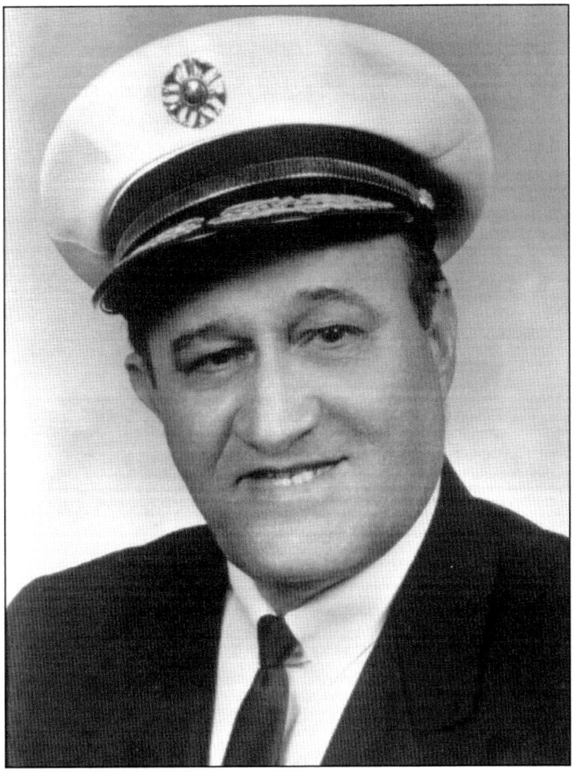

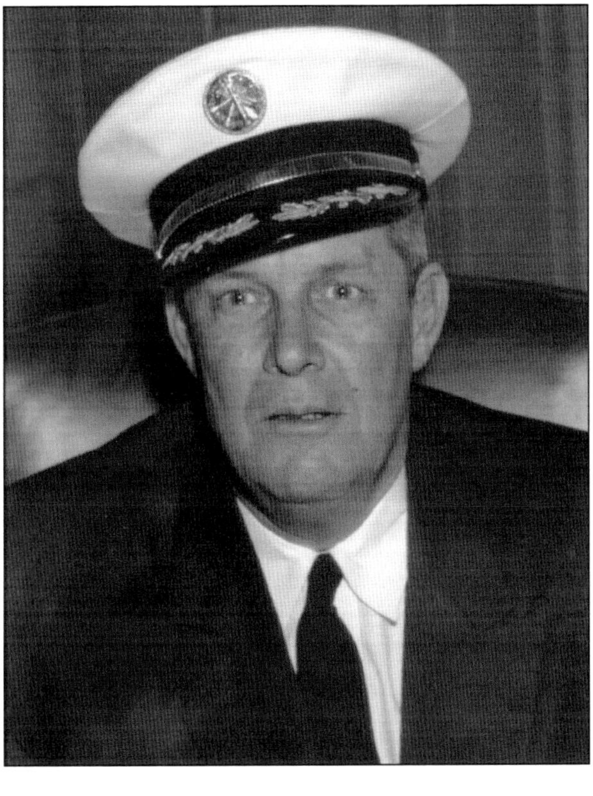

District Chief Charlie Bullock was named fire chief in 1958. Under Bullock, a record was set for the number of calls in a 24-hour period, on September 11, 1961. Firefighters answered 170 calls, due primarily to Hurricane Carla. In 1963, a third shift was created, reducing the average firefighter's week from 60 hours to 56 hours, involving three 10-hour days, three 14-hour nights, and then three full days off.

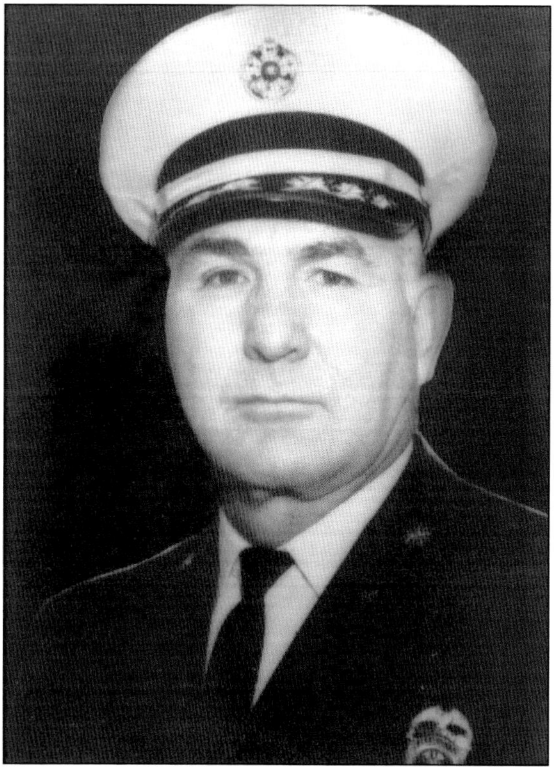

Assistant Chief William O. Hunter was named fire chief in 1964 after Mayor Louie Welch was elected. Chief Hunter instituted a home inspection program, saw the city council raise salaries of firefighters to spur recruitment, and had two snorkels and a small fireboat put into operation. In 1967, civil unrest in Houston resulted in firebombed businesses, spurring the halt of Local 341's Ladies Auxiliary for safety concerns.

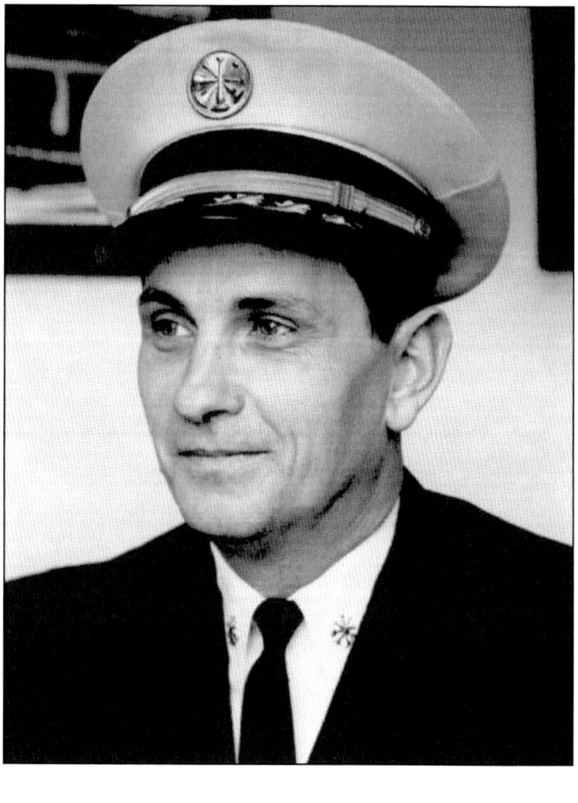

Assistant Chief Clarence Cook was promoted after Chief Hunter retired. Under Cook, a new four-story police/fire headquarters building opened, the city outlawed fireworks inside the city limits, emergency medical service was added, and fire apparatus changed from red to white. In addition, a new motor-repair and maintenance shop, international airport, and fire-training facility opened. Chief Cook resigned in 1973 amid allegations of wrongdoing; he was later acquitted.

Jim Honea was named acting fire chief after Chief Cook's resignation in 1973. That October, the firefighters' union fought for and won equal pay rates between firefighters and police officers. Also, the city council approved the first comprehensive fire code in November, which had been drawn up by fire-prevention personnel. Honea retired at the end of 1973. Assistant Fire Chief Fred A. Hooker was named acting fire chief for a few months.

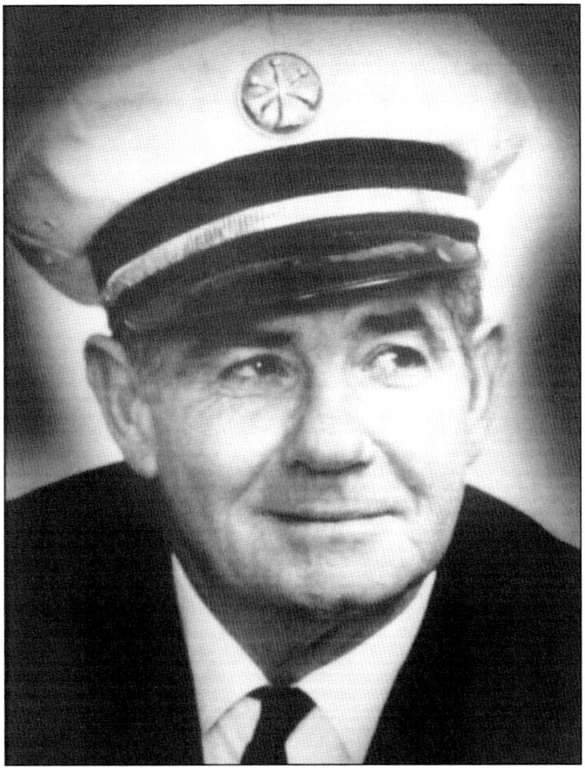

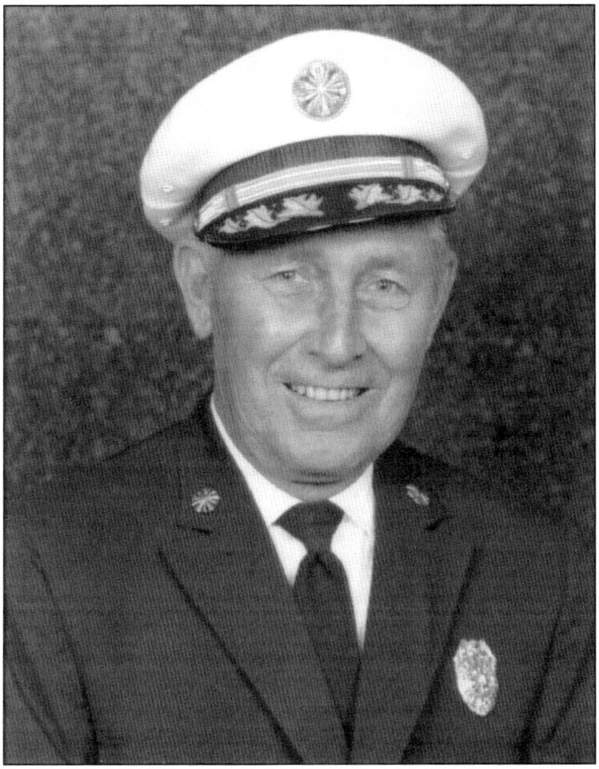

District Chief John Little was selected by Mayor Fred Hofheinz in 1974. Following debate and a nonofficial strike over the department's promotion policy for EMTs and paramedics, a bid for collective bargaining was soundly rejected. Linda Honeycutt began her training in the department in March 1975, becoming Houston's first female firefighter. A new fire alarm building opened in December at Preston and Bagby Streets. Chief Little retired in January 1976.

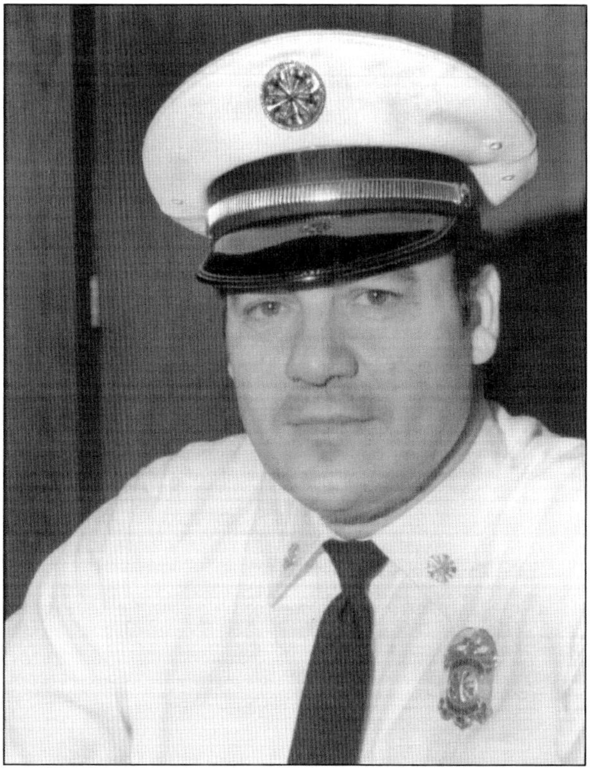

Sr. Capt. Joe Perino, a former firefighters' union president, was named fire chief in February 1976, following Little's retirement. The Uniform Fire Information Reporting System, a computer program for recording and reporting fire information, was adopted in his first year. Removal of street fire-alarm boxes began the same year; only about 100 of the 832 fire-alarm boxes remained by the following year.

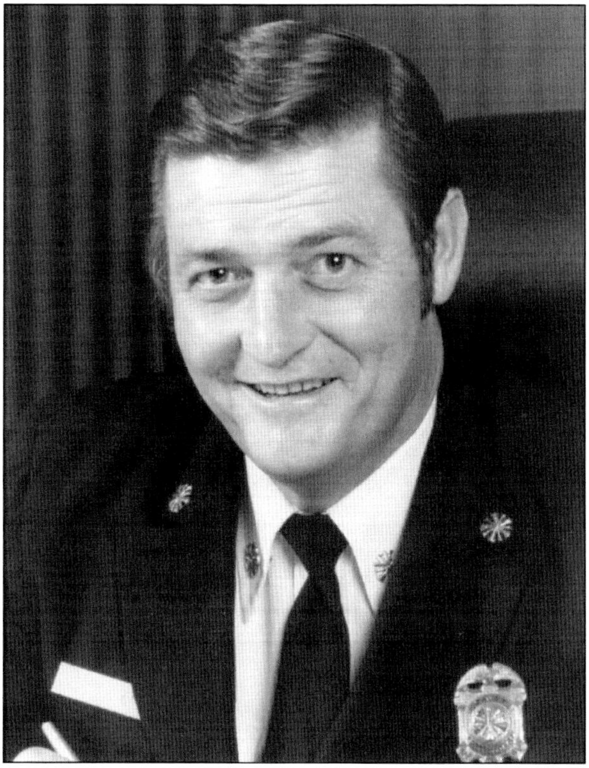

District Chief V.E. Rogers was named fire chief in 1978. Rogers revised dispatch policy to include an extra engine and ladder truck to all large apartment-complex fires, initiated the Houston Hazardous Materials Response Team, and saw workweeks reduced. During his tenure, old Station 7 was preserved as a museum, seven rescue and salvage trucks were put into service, and a new computer-assisted dispatch system was completed. Chief Rogers retired in May 1984.

After an interim stint by Assistant Chief Raymon Harrison, Robert L. Swartout, fire chief of Seattle Fire Department, was named chief in December 1984 following a change in city ordinance permitting the institution of a fire chief from outside HFD. Swartout quickly discovered that Houston politics were not for him; within four months, he resigned and returned to Seattle.

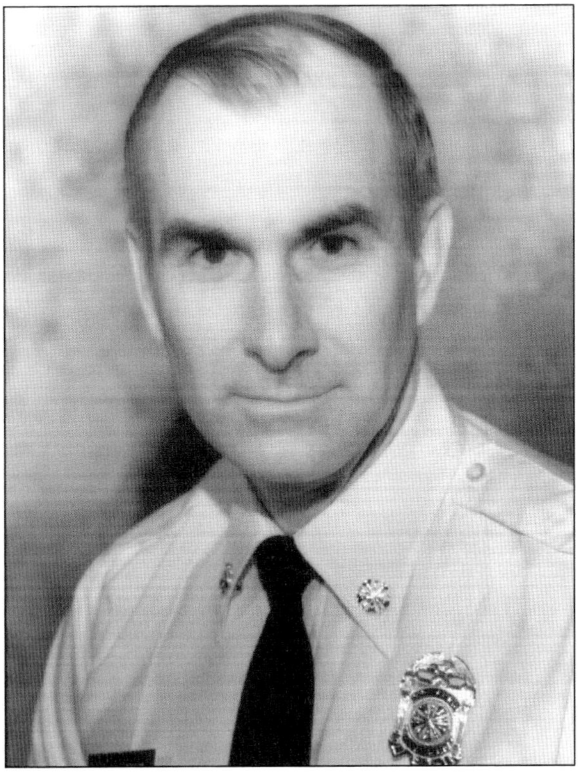

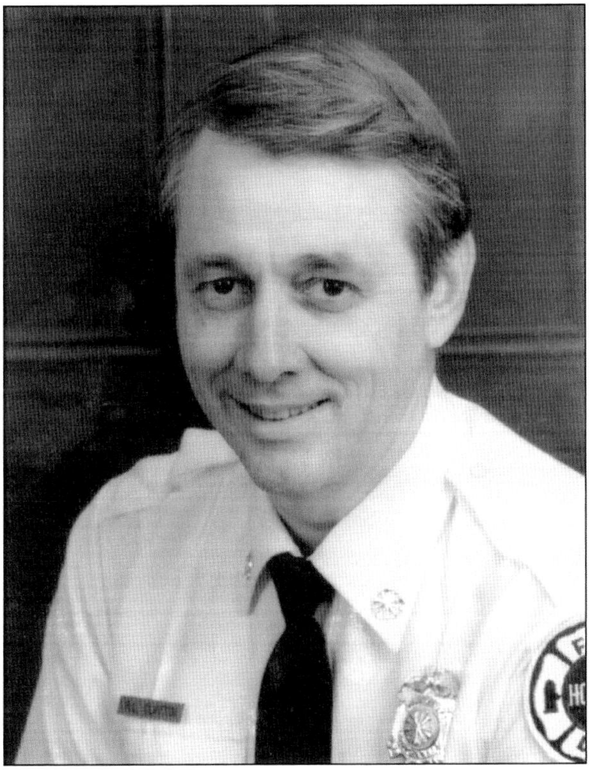

Deputy Chief Robert Clayton was quickly picked to replace Swartout. He faced immediate departmental changes. These included the elimination of six fire districts, the closing of divisions, the shifting of positions, and a headquarters move. In contrast, the department acquired 68 apparatus, implemented 911, and improved basic command procedures. In December 1992, the mayor was defeated, and Chief Clayton's tenure came to an end.

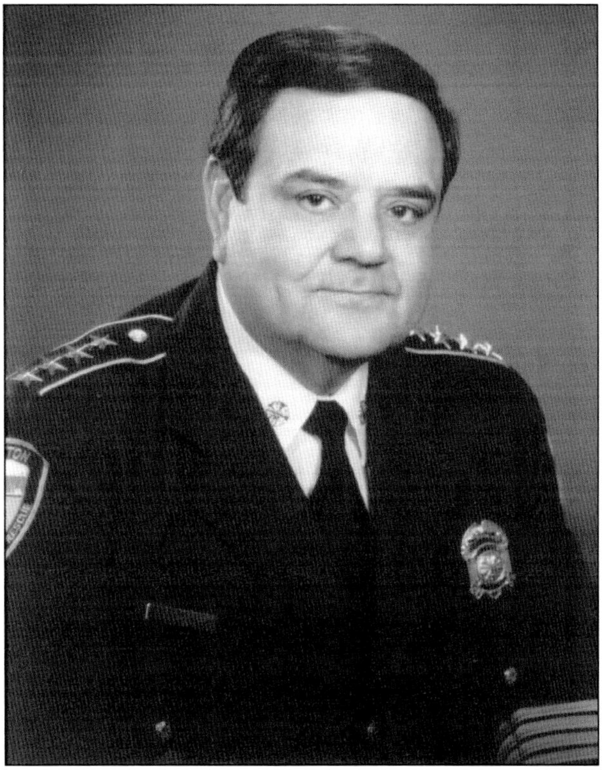

Fire Marshal Eddie Corral was named the first Hispanic fire chief in 1992 by the new mayor. Training of new recruits returned to the Fire Training Academy from a community college, and 80 female firefighters were hired. Only three women were in the department when his tenure began. Response to reported fires in high-rise buildings was increased in February, and Chief Corral began the HFD's renowned Honor Guard in 1992.

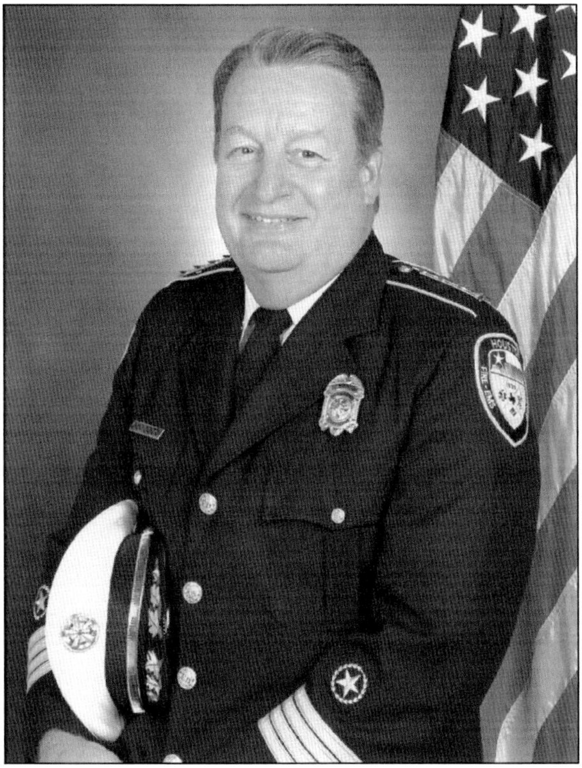

District Chief Lester Tyra started as fire chief in March 1998. Chief Tyra immediately conducted an evaluation of department operations, resulting in year-round recruitment; full training academy staffing; enhancement of drills, training, and education; purchasing of new apparatus, thermal imagers, and cellphones; improved fleet maintenance and fire prevention; and development of an officer training program. Tyra stepped down in October 2000, returning to his district chief position.

Assistant Chief "Chris" Connealy became chief in 2000. Connealy was instrumental in making the department the largest accredited fire department in the world in 2001. Paramedic squads equipped with advanced life support equipment were instituted, and the Get Alarmed Houston home smoke-detector program was started. Chief Connealy retired in January 2004, and Assistant Chief Hector Trevino served temporarily.

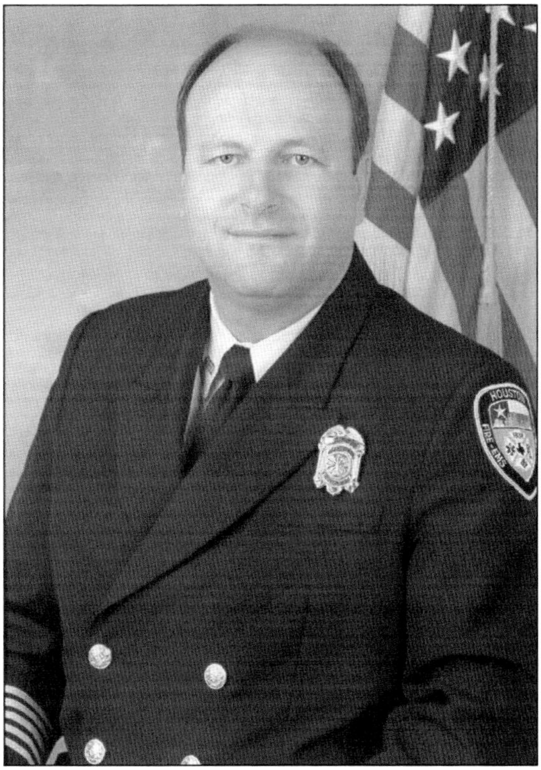

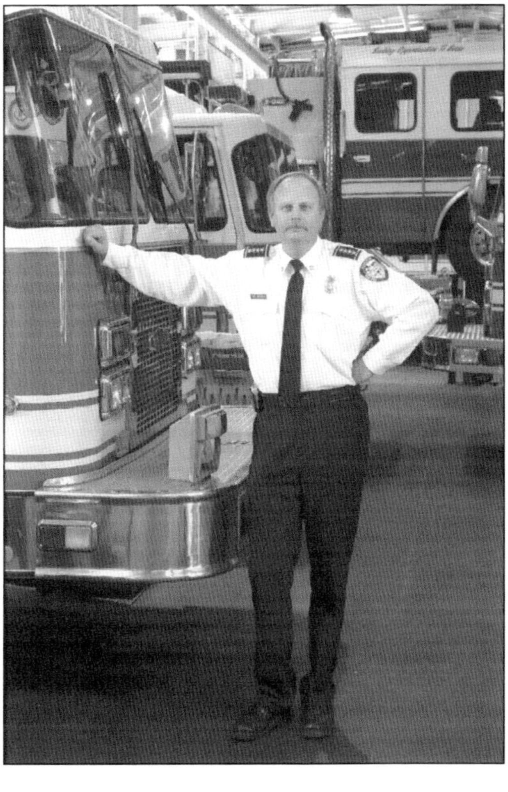

District Chief Phil Boriskie, a 22-year veteran, was picked as fire chief in May 2004 after an extensive nationwide search. Boriskie pledged renewed efforts at firefighter safety and negotiated a contract with the union, something firefighters had been without for several years. Hurricane Katrina saw firefighters pressed into round-the-clock service in early September 2005. They triaged refugees who escaped flooded New Orleans for Houston.

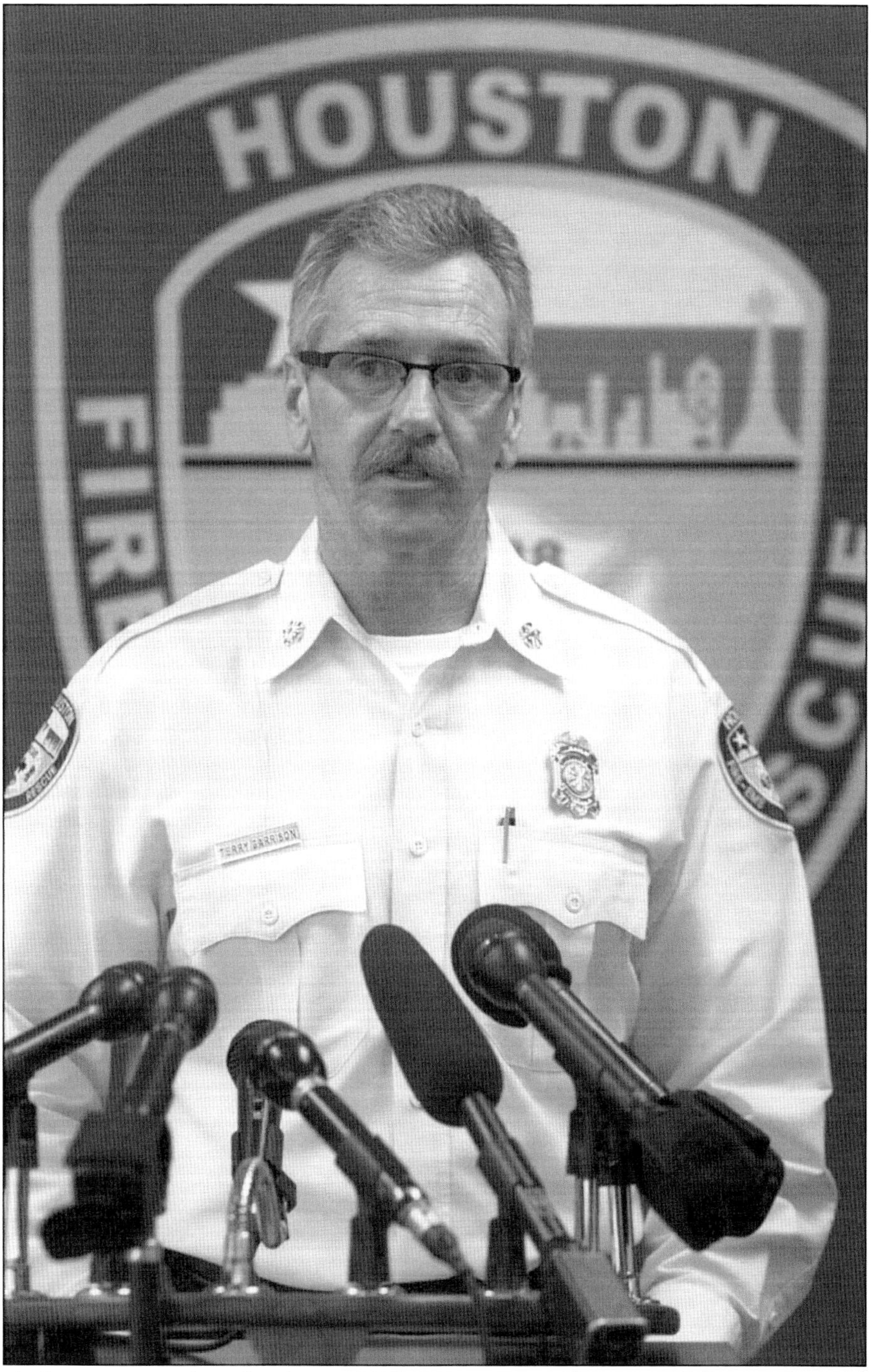

Terry A. Garrison was selected in August 2010 after serving in Phoenix and Oceanside, California. Garrison installed the most diverse command staff in the department's history. He also oversaw the response to the department's worst casualty event, a May 2013 fire in which four firefighters were killed in the line of duty and five others injured. He also managed a budget shortfall resulting in staffing and vehicle cuts.

Six

LINE OF DUTY DEATHS

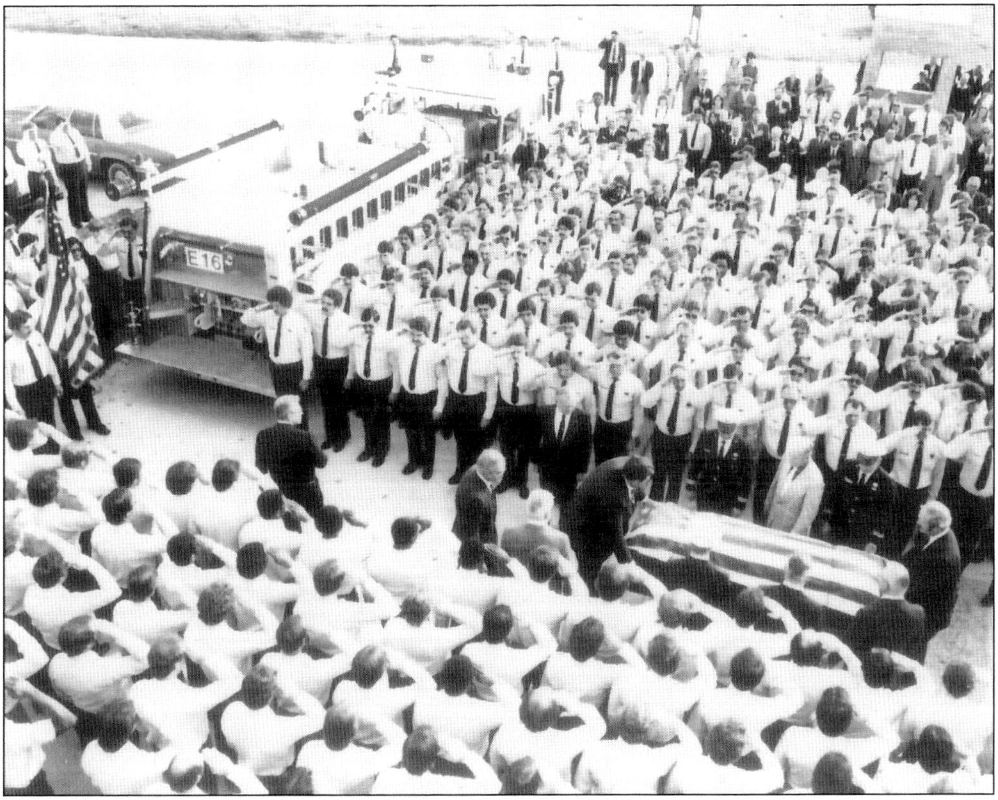

The first recorded line of duty death (LODD) in the Houston Fire Department occurred on April 27, 1889. Since that date, 68 firefighters have perished serving the citizens of Houston. The leading cause of firefighter fatalities in the nation is heart attack, followed by trauma, motor-vehicle collisions, asphyxia, and burns. When a fire department loses a member in a LODD, it will commonly fly its flag at half-staff, and members will cover their badges with a mourning band. Seen here is the funeral of District Chief Lonnie Franklin.

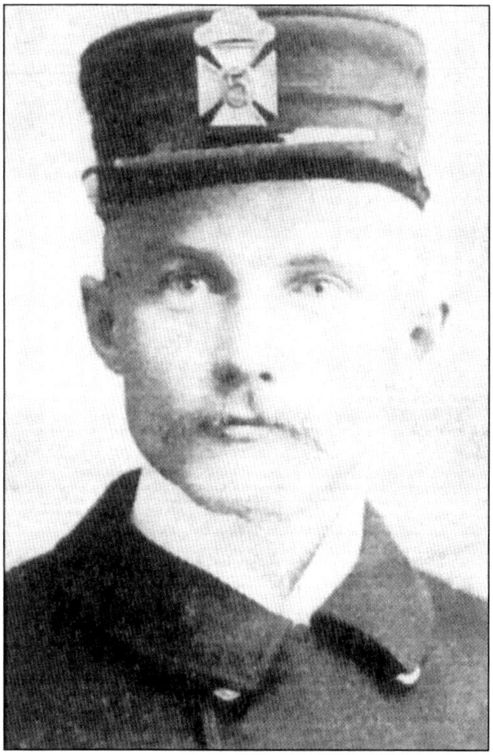

Hose Wagon No. 5 was en route to a fire on April 7, 1899, when it encountered a blind crossing along railroad tracks. Men riding the rig feared an approaching train was about to hit them, and they jumped. Pipe & Ladderman Emil "Ed" Thompson, the last man to jump, fell backward into the path of the train and was killed instantly.

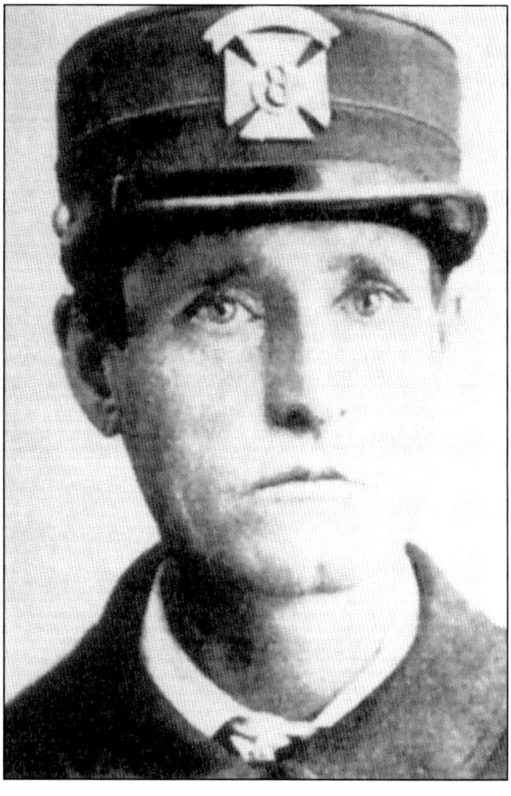

Driver Frank W. Medlenka, Steamer No. 8, was killed on November 23, 1901, in an accident while responding to a fire at the corner of McKinney and Chartres Streets. The steamer that he was driving tipped over as it crossed over loose planks covering new streetcar tracks being installed at McKinney and Jackson Streets. Medlenka, who entered the department on September 21, 1895, was crushed under the overturned steamer.

Pipe & Ladderman Clifford W. Johnson was killed while driving the assistant fire chief to a blaze at McKinney and Chartres Streets. Johnson was traveling up Crawford Street in front of Station 8 (seen here around 1905) when he plowed into the rear of a slow-moving vehicle without lights. He died early the next day on June 13, 1902.

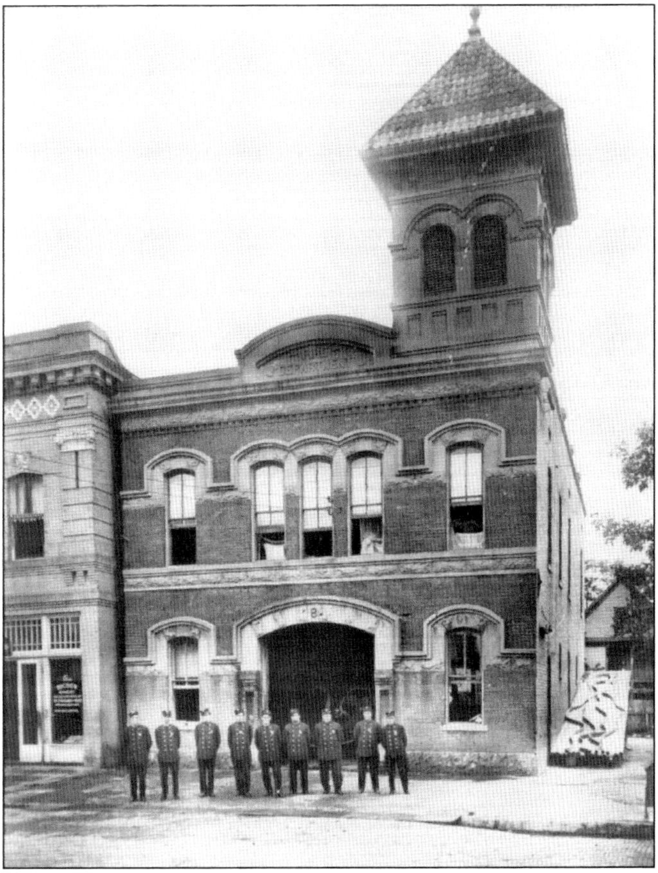

Pipe & Ladderman Barney C. Moffatt of Station 7 was killed fighting a fire at Mercantile Grain Company, 1215 Hardy Street, on April 26, 1908. He and his captain were overhauling in the main warehouse atop high piles of sacks of grain. Without warning, the floor gave way, and Moffatt plunged to his death in the basement. His captain was able to save himself by grabbing onto a ceiling beam.

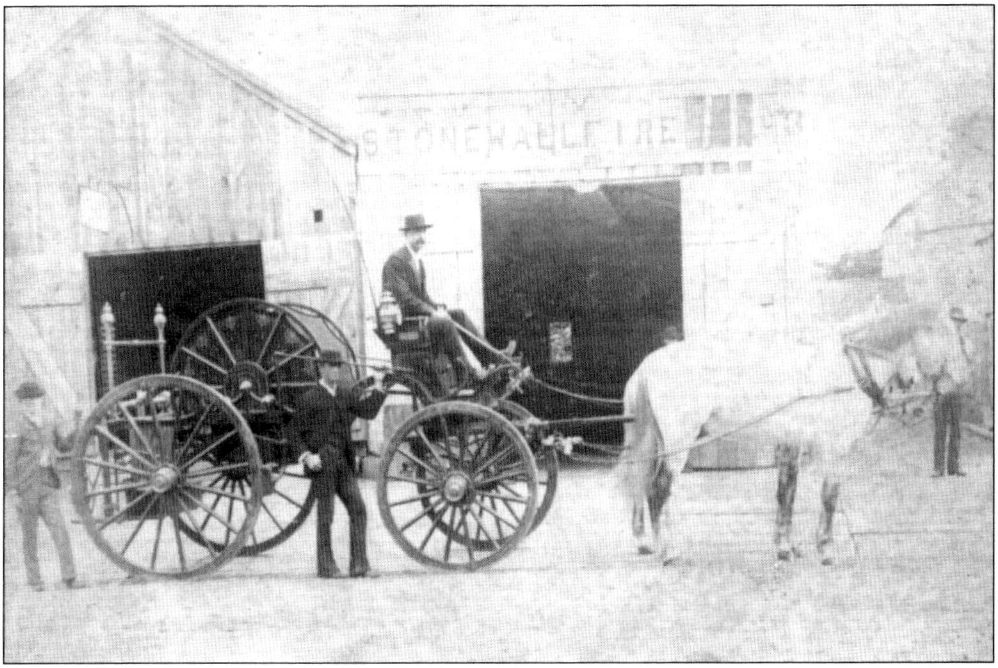

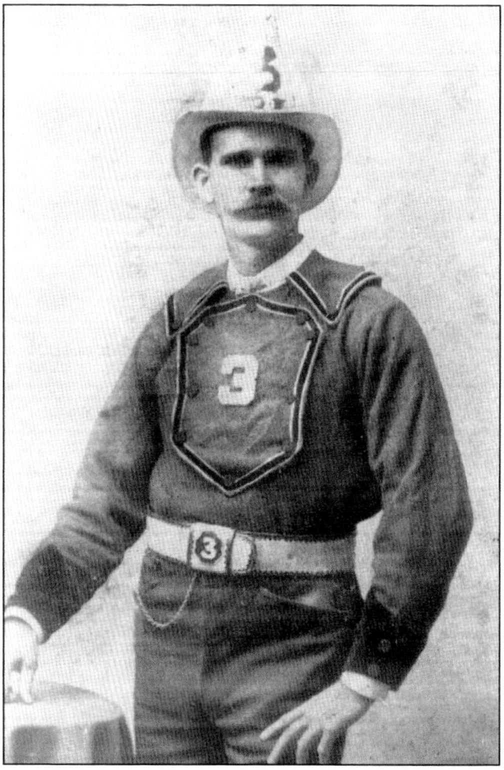

Fire Chief Thomas O'Leary lost his life while conducting operations at a blaze involving a fireworks-laden boxcar on November 22, 1908. An explosion occurred, and Chief O'Leary took the full blast of a second explosion, believed to be from one or more unexploded shells in the boxcar. He never recovered from his massive injuries and died the following month, on December 24.

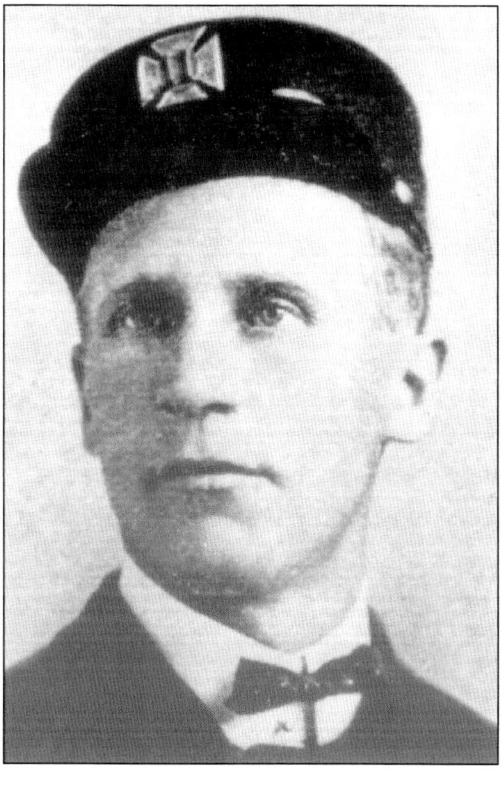

Capt. Herman "Dutch" Wagner was critically injured on May 19, 1912, at the Stower's Furniture fire in the six-story Mason Building on Main Street. A heavy timber struck Wagner as he and his crew from Station 8 battled the inferno that destroyed Stower's and three adjacent buildings. Wagner died of his injuries in St. Joseph Hospital six days later, on May 25.

Pipe & Ladderman Clifford Weise was assigned to the supply wagon at Station 3 on July 14, 1913. Weise was headed back to the station after delivering supplies and was hit by a work-crew train at the Johnson Street crossing. The lead car splintered the supply wagon and seriously injured both Weise and the horses pulling the wagon. Weise died two hours later at St. Joseph Hospital.

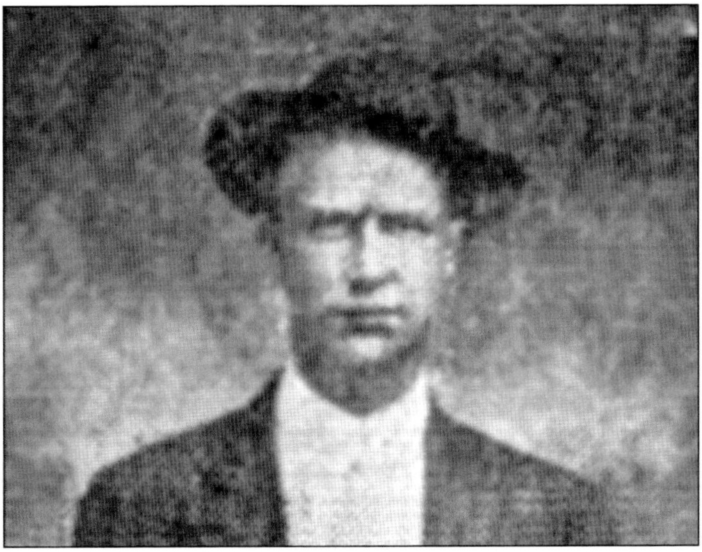

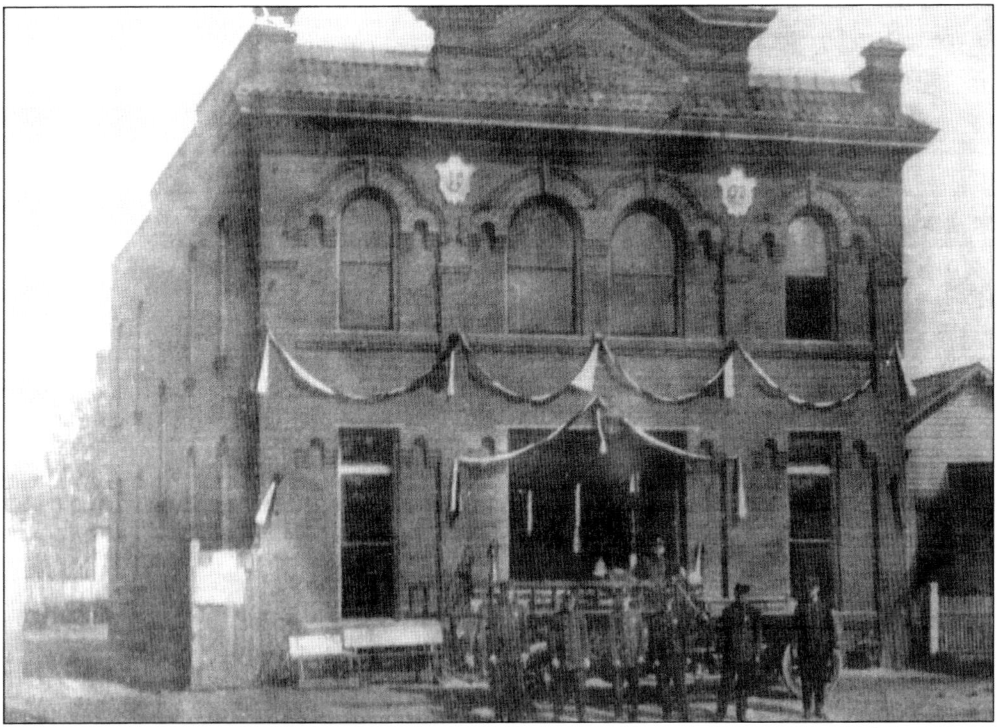

Pipe & Ladderman Patrick O'Hara, Station 6, was killed en route to a fire on September 21, 1913. He was thrown as his engine maneuvered through an offset on Sabine Street where it crossed Lubbock Street. A rear wheel struck a hole, and O'Hara was thrown into a telephone pole. At the time of the incident, he was donning his quick-hitch coat and did not have a firm grip on the handrail. O'Hara was killed instantly.

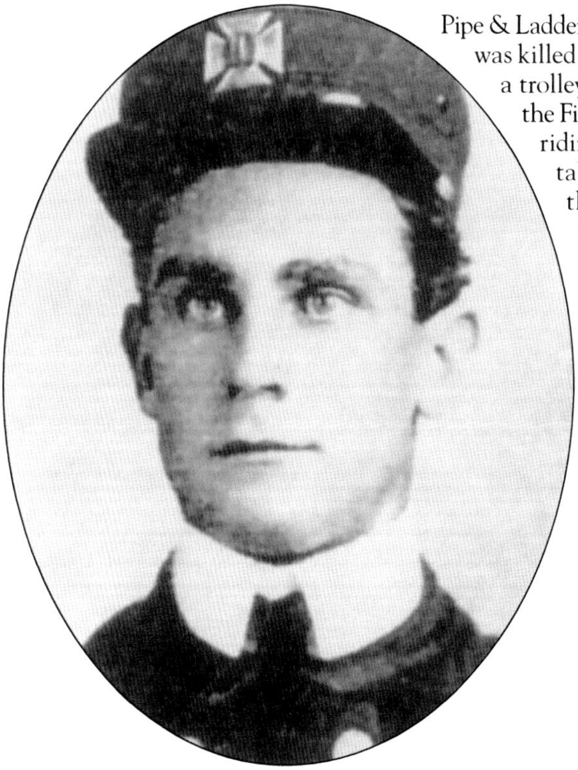

Pipe & Ladderman Robert Glass of Central Fire Station was killed when Service Truck No. 1 slammed into a trolley car as the firefighters raced to a fire in the Fifth Ward. Glass and two other firefighters, riding the side of the rig, were injured and taken to St. Joseph Hospital. Glass died the following morning, on December 1, 1913. He had entered the department in 1902.

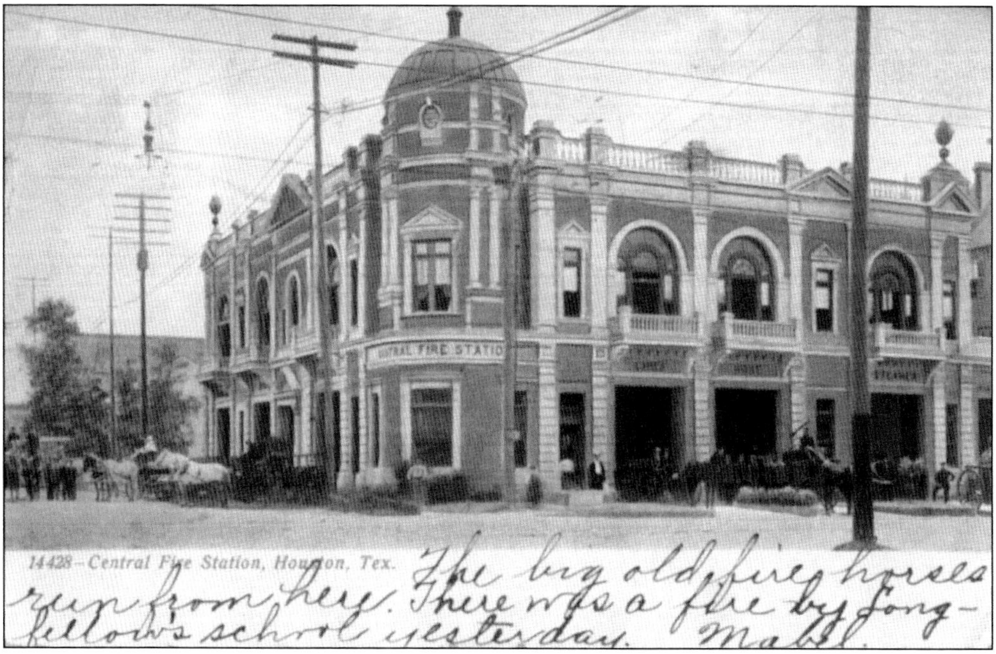

Pipe & Ladderman Harry Dodd of Central Fire Station (seen here) was killed in a collision at the intersection of Dallas and Milam Streets on July 26, 1923. He was riding on Ladder No. 1 headed to a fire at 1315 Shaw. The ladder truck struck an automobile at the intersection and careened head-on into a trolley car that had stopped for the fire truck. Dodd died instantly.

Pipe & Ladderman H.A. Harris, of Engine No. 10 (pictured here), was killed on February 14, 1921, battling a blaze in the Texas Lamp & Oil Company at Baker and Pine Streets. He was manning a hose line in the alley when a wall of the three-story building collapsed. Harris was trapped under the debris and died of suffocation from the smoke.

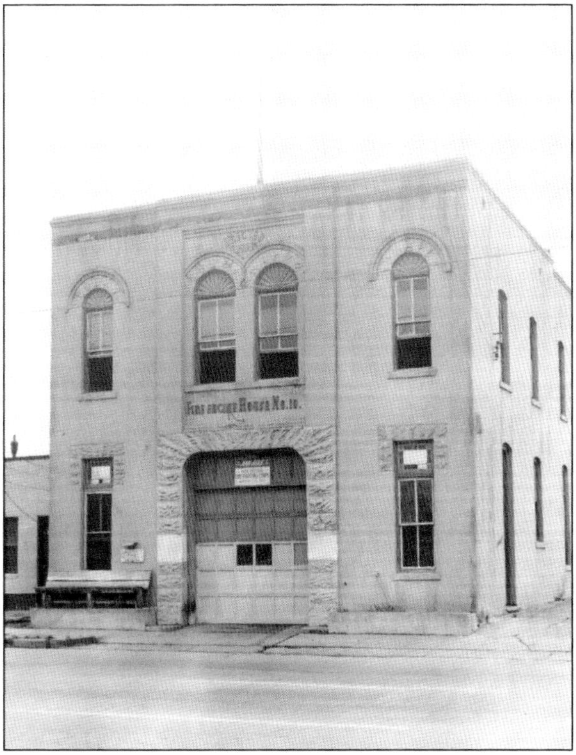

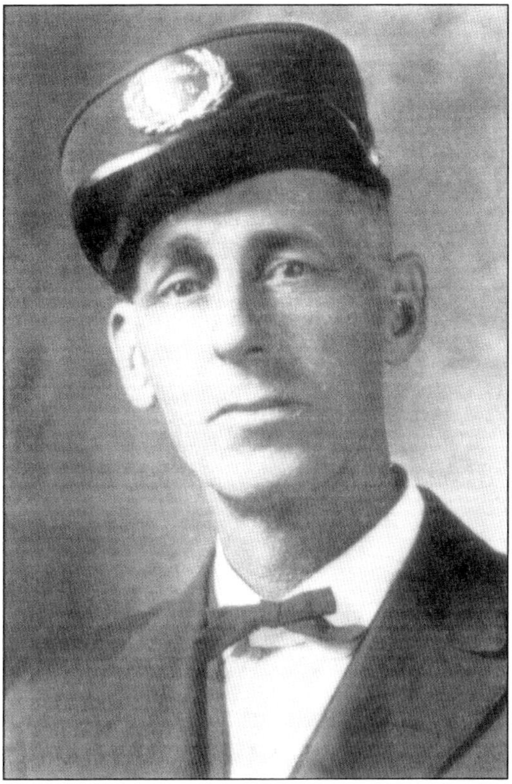

First Assistant Fire Chief "Bob" Boyd of Central Fire Station was killed on August 4, 1923, en route to a fire on Shrimp Alley Street. A raging fire was in progress, and he hopped aboard Chemical No. 1 as it was responding to the alarm. The driver of the chemical rig swerved to avoid a collision with an automobile and slammed into a telephone pole.

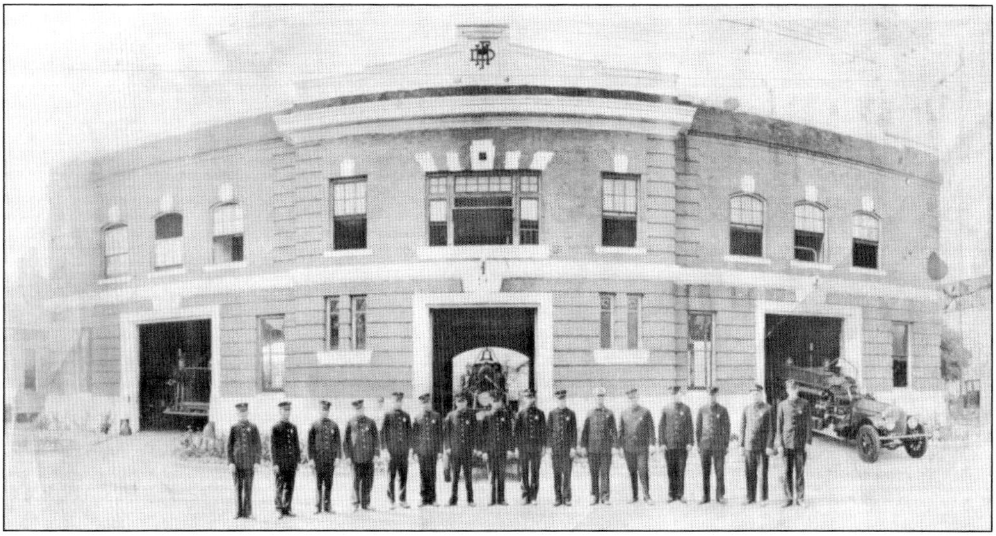

Pipe & Ladderman Harold Denny was killed on May 1, 1924, en route to a fire at the Bluefront Boarding House at Preston and Milam Streets. He was riding the tailboard of Engine No. 9 (the crew of which is pictured here in the 1920s) when it was struck by a streetcar as the engine exited the North Main Street viaduct. Denny, not on duty, had dropped by the station when the alarm sounded. He jumped aboard to make the fire.

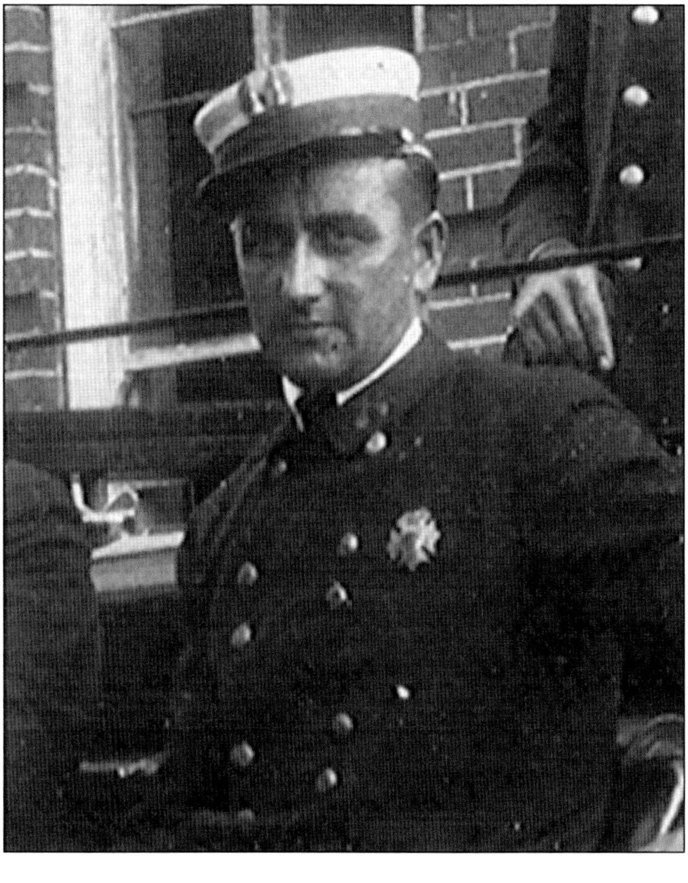

Jr. Capt. George L. Bishop, Station 4, was killed during a house fire on Grant Street. Bishop led his crew up the rear outside stairs. After reaching the second-floor porch, the chimney collapsed, tearing the porch loose and sending it to the ground. Bishop took the brunt of the showering bricks and debris. He was taken to St. Joseph Hospital, where he died the following morning, March 1, 1926.

Claude Edmonds, acting captain at Central Fire Station, was killed on August 19, 1926, in a training exercise adjacent to the station. The firefighters were climbing an 85-foot aerial ladder and returning on the underside of the ladder. The ladder broke, and Edmonds fell just after he had climbed over the tip of the ladder. He died from a crushed skull and chest and a broken neck and back.

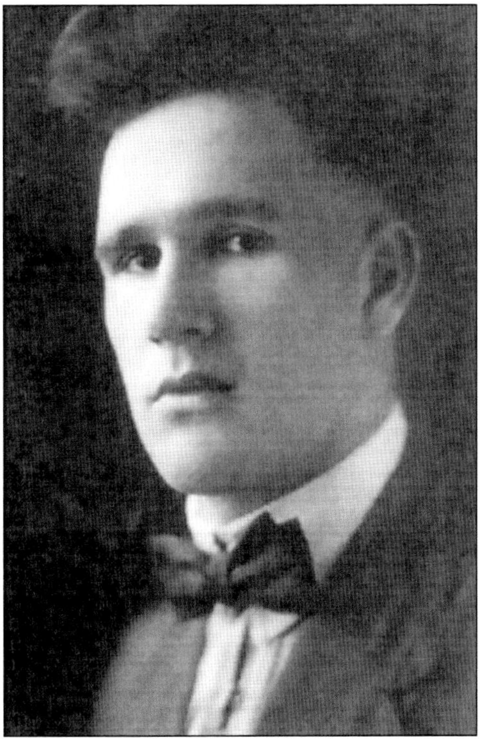

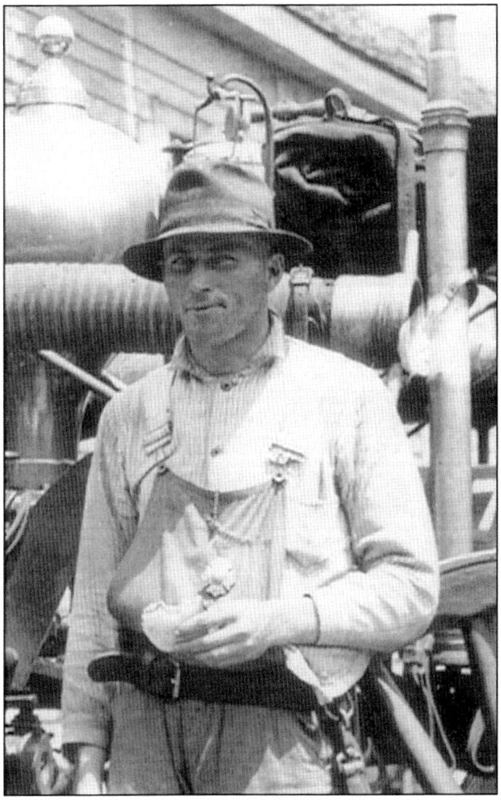

Station 18 Pipe & Ladderman James "Mag" Smith was killed on February 9, 1929. Smith caught a fire hydrant at a fire on Congress Avenue and had the hose wrapped securely around the plug. Engine 10 proceeded on to the fire, laying out the hose, when the hose suddenly hung up and pulled from Smith's grip. Unwinding from the wrap-around, the coupling struck Smith in the head, killing him instantly.

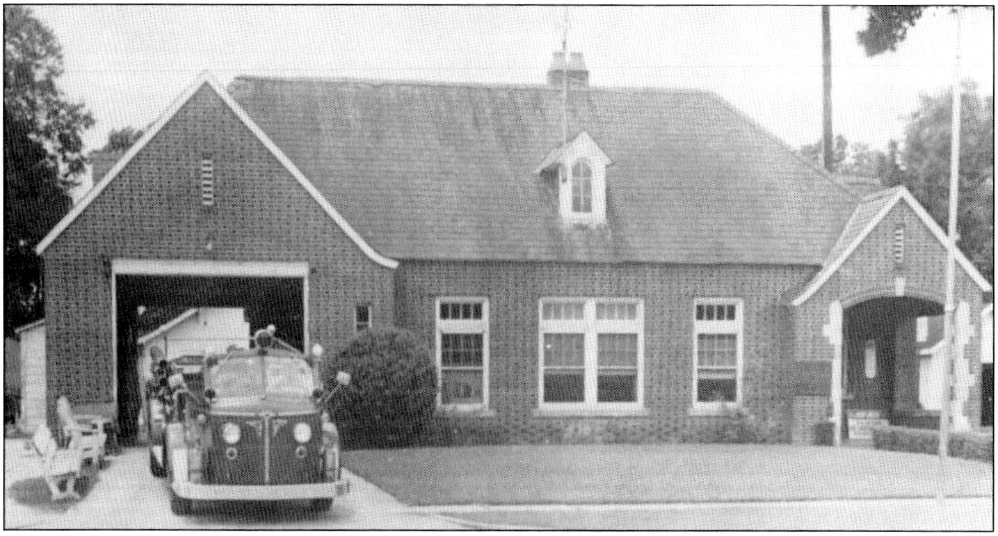

Pipe & Ladderman Rudolph Hendricks, of Engine No. 25 (pictured here), was electrocuted at a fire on June 22, 1929, at 3252 Reeves Street. He was operating a nozzle at the rear of the fully involved house when the overhead electrical drop burned loose and fell. The falling wire encircled him, and he died despite the immediate administration of artificial respiration by fellow firefighters. Hendricks had entered the department on January 4, 1928.

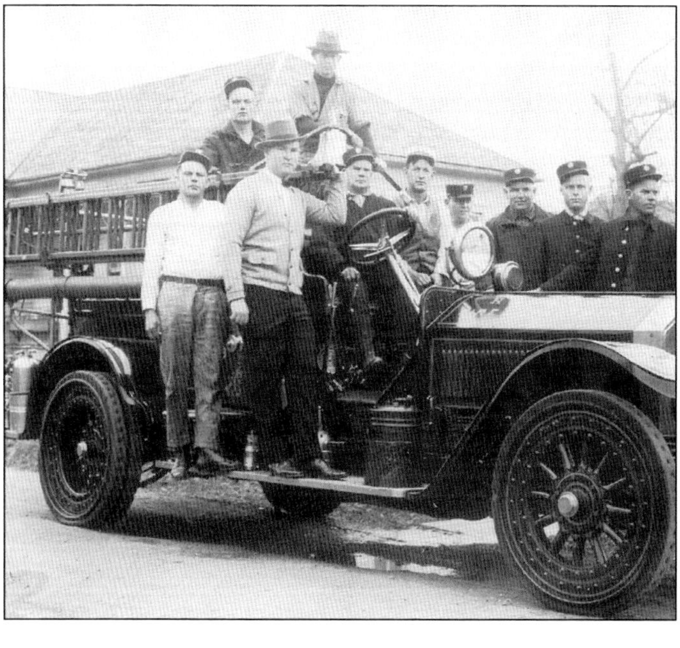

Pipe & Laddermen Edgar Grant (far right) and Harry "Red" Oxford (on running board at right) were killed on July 2, 1929, when a train broadsided Engine No. 18 on Telephone Road at Lombardy Street. They were responding to a house fire at 2426 Wilshire Street. Engine No. 18 was knocked from the railroad crossing, overturned, and completely destroyed, killing Grant instantly. Oxford died the following day, and Captain John Little Sr. (fifth from right) died on July 11, 1929, from injuries suffered during the collision.

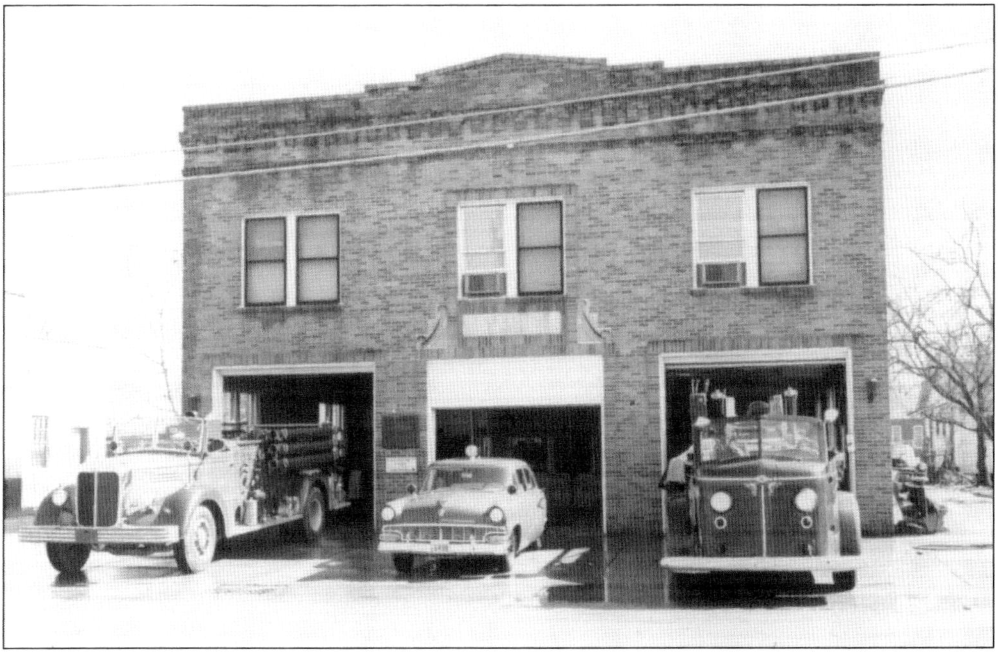

Pipe & Ladderman Aaron O'Brien was killed on December 20, 1929, in an accident involving Engine No. 19 (pictured here) and a Southern Pacific train. The engine was responding to a fire on Quitman Street near Clark Street when it was struck by the train at the railroad crossing just north of the station, which was located on Gregg Street. Off-duty firefighter William M. Ward died of injuries from the same accident.

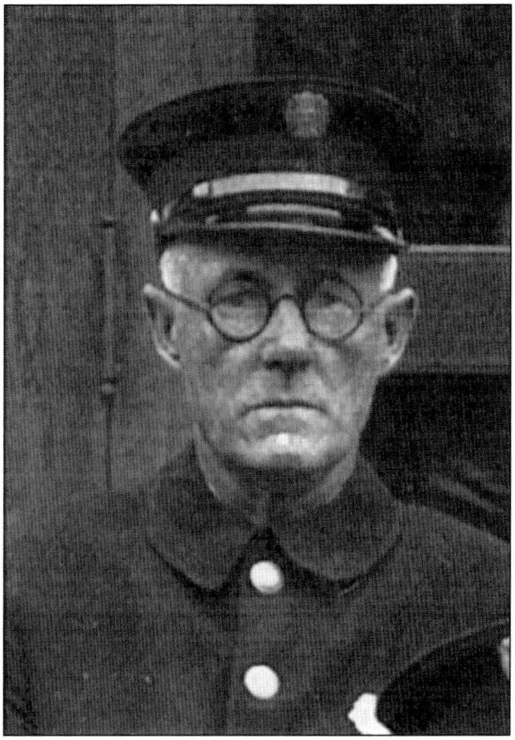

Pipe & Ladderman Charles Rusk was killed on June 8, 1932, in a collision between Ladder No. 14 and a streetcar at Seventeenth and Yale Streets. The ladder truck, responding to a house fire, slammed into the streetcar that was crossing Seventeenth Street in front of the rig. Rusk died at Heights Clinic later that day. He had entered the department on April 24, 1918.

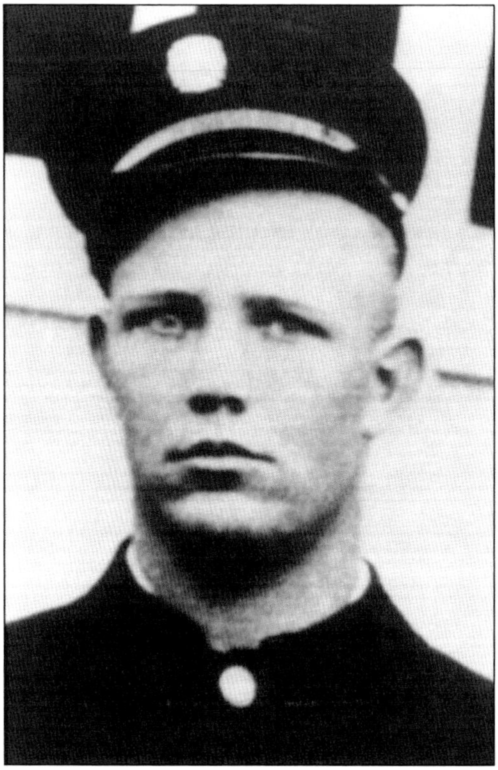

Esker Rogers, pipe & ladderman at Station 17, was filling in at Station 24 on August 22, 1934, when a box alarm came in. Engine No. 24 responded and was en route on Leeland Street when a streetcar pulled across Jackson Street. To avoid a collision, the chauffeur turned the engine sharply, but the skidding rig smashed sideways into the streetcar. Rogers was crushed between the engine and streetcar and killed instantly.

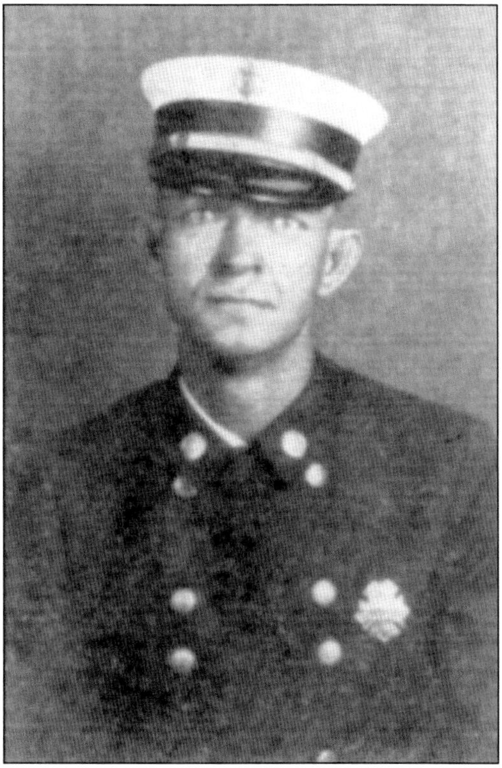

District Chief Alfred Schutze was killed in a nighttime accident on August 21, 1938, responding to a garage apartment fire at 2012 Bolsover Street. His chemical truck, housed at Station 7, struck the esplanade in the 5100 block of Main Street and overturned. An amber light, warning of the esplanade, had burned out.

Chauffeur Walter W. "Molly" Walker of Station 2 died a year and a half after he was injured in an accident on December 6, 1938. He was driving Engine No. 2, as a reserve for Engine No. 17, to a fire and struck a bus at the intersection of Harrisburg Boulevard and Sampson Street. Walker never recovered from the injuries and died on July 1, 1940.

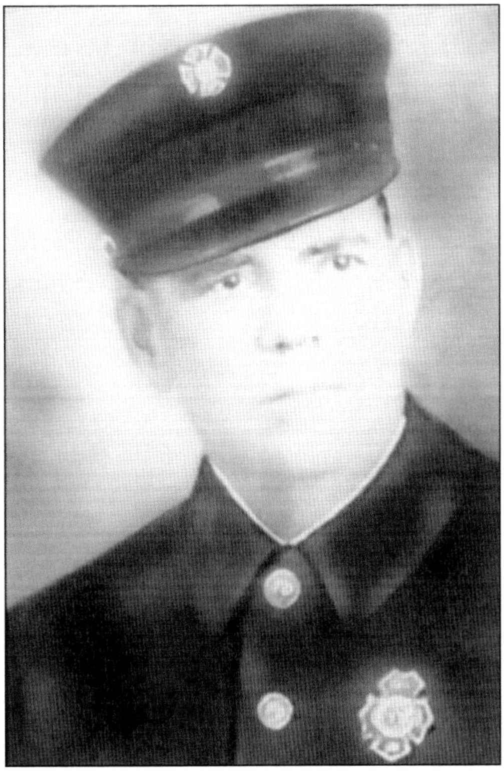

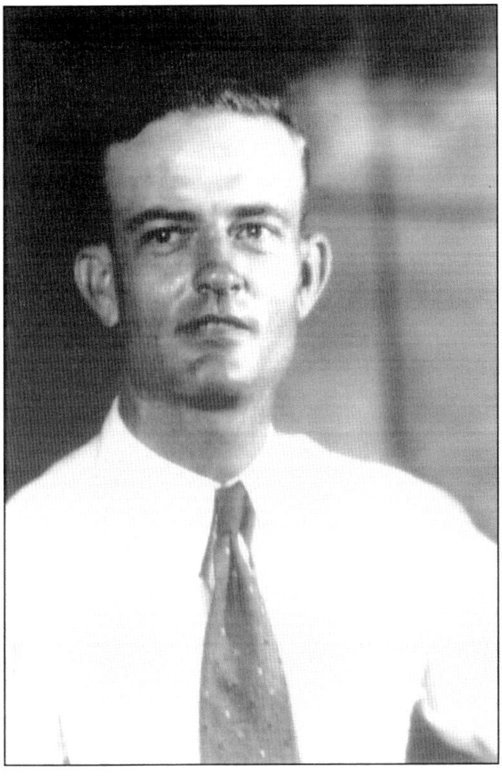

Pipe & Ladderman Horrace Gassoway, assigned to Station 10, was killed in an accident on August 6, 1941, en route to a supermarket fire at Preston Street and Washington Avenue. He was filling in on Service Truck No. 1. On the way to the fire, Engine No. 2 swung in behind Truck 1 at Preston and Smith Streets and could not stop. Gassoway was thrown from the truck in the collision.

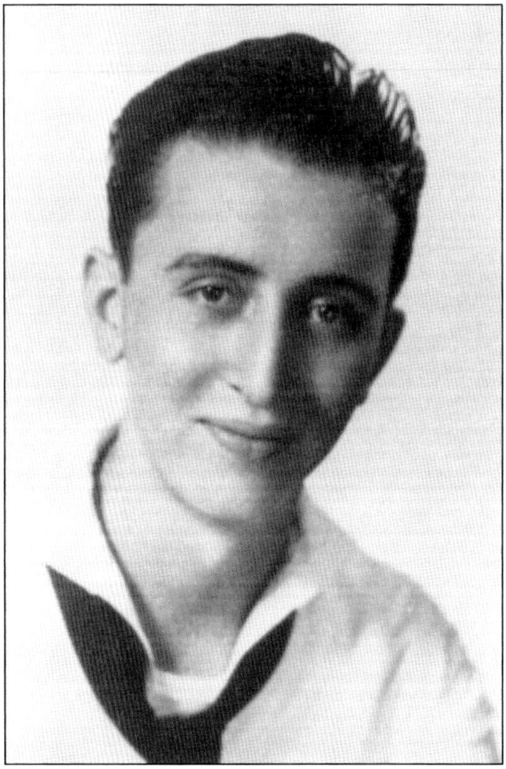

Pipe & Ladderman James Edward Lowth of Station 26 was killed instantly on March 31, 1947, in a collision en route to a fire at Eastern States Refinery. Engine No. 26 collided with a Southern Pacific Railroad train at Broadway and Bowie Street. Lowth had recently served a 41-month stint with the US Navy in World War II.

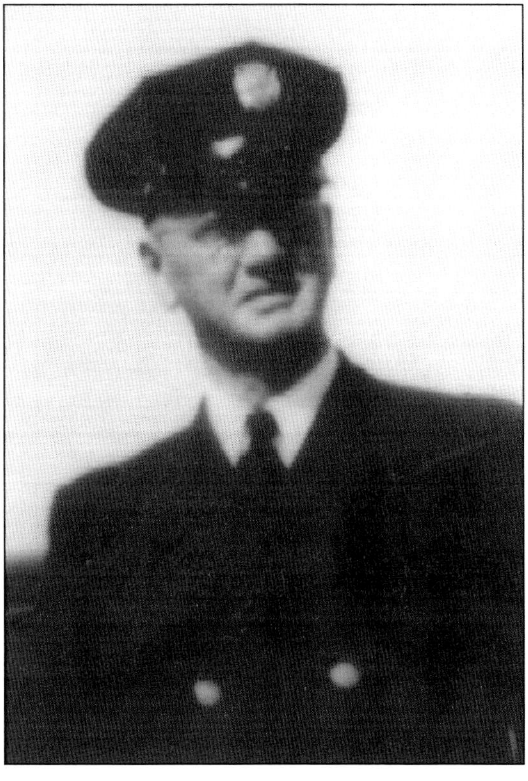

Pipe & Ladderman John "Red" Adams, from Fire Station 18, died from a heart attack on March 7, 1948, at a house fire at 2850 Lidstone Street. The acting chauffeur, Adams collapsed while pumping on Engine No. 18. The crew was dispatched to contain the blaze, which turned out to be a small fire in a closet. The Harris County Emergency Corps arrived, but attempts to revive Adams failed.

Assistant Fire Chief Vernon Dorsett of Central Fire Station was killed on January 21, 1949, in an accident at the intersection of Harrisburg Boulevard and Linwood Street. Chief Dorsett responded to an alarm at the Hub Lumber Company at 5519 Lawndale Street. En route, his car hit wet pavement and collided with a car with a mother and child inside. They sustained minor injuries, but Dorsett was killed instantly.

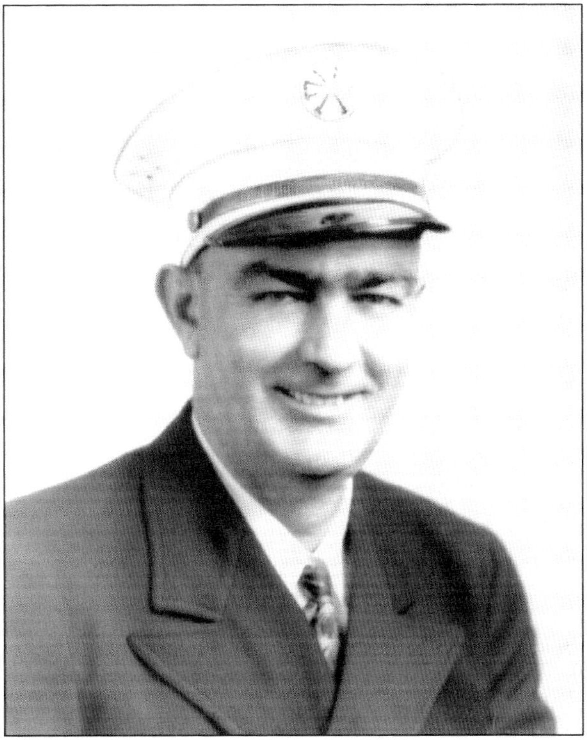

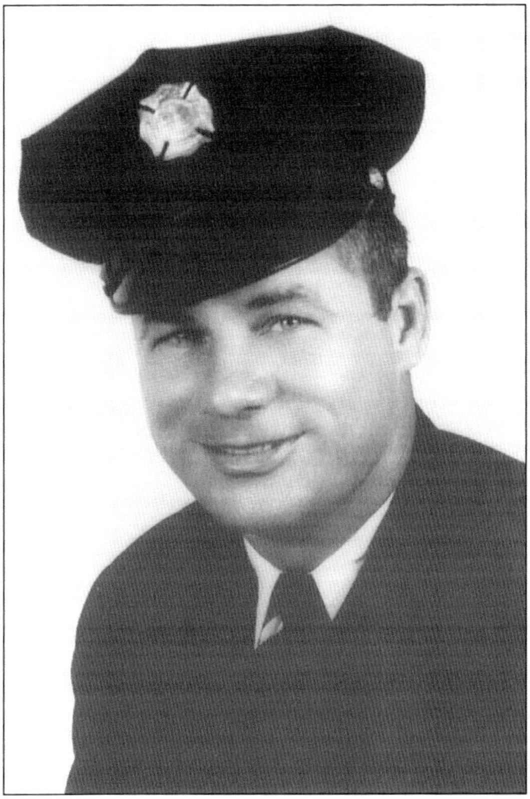

A collision between the No. 7 chief's car and a pickup truck on April 14, 1949, killed Chauffeur Robert "Bobby" Milburn. He was responding to a box alarm when the accident happened at the intersection of McGowen and Fannin Streets. Both the district chief and the driver of the pickup truck were seriously injured. Milburn was thrown from the car. He had entered the department on September 4, 1933.

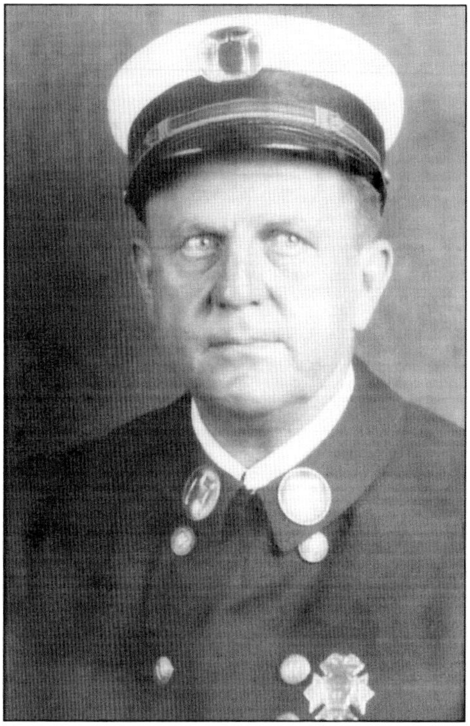

On January 31, 1951, Rufus "Bullet" Templet suffered a heart attack at the Alaskan Fur Factory on Main and Texas Streets. A rare cold snap brought temperatures to an extreme low, and the icy roads and thick smoke created dangerous conditions for the firefighters. Templet, a captain at Station 5, was rushed to Veterans Hospital, where he died two months later, on March 12.

On January 31, 1951, Chauffeur Joseph "Bull" Connor of Station 1 suffered a fatal heart attack, in part due to abnormally icy conditions. Connor was setting up a heavy stream on the water tower at a fire in the Alaskan Fur Factory at Main and Texas Streets (seen here). Efforts to revive him at the scene proved futile. Another firefighter, Rufus Templet, also suffered a heart attack during the fire.

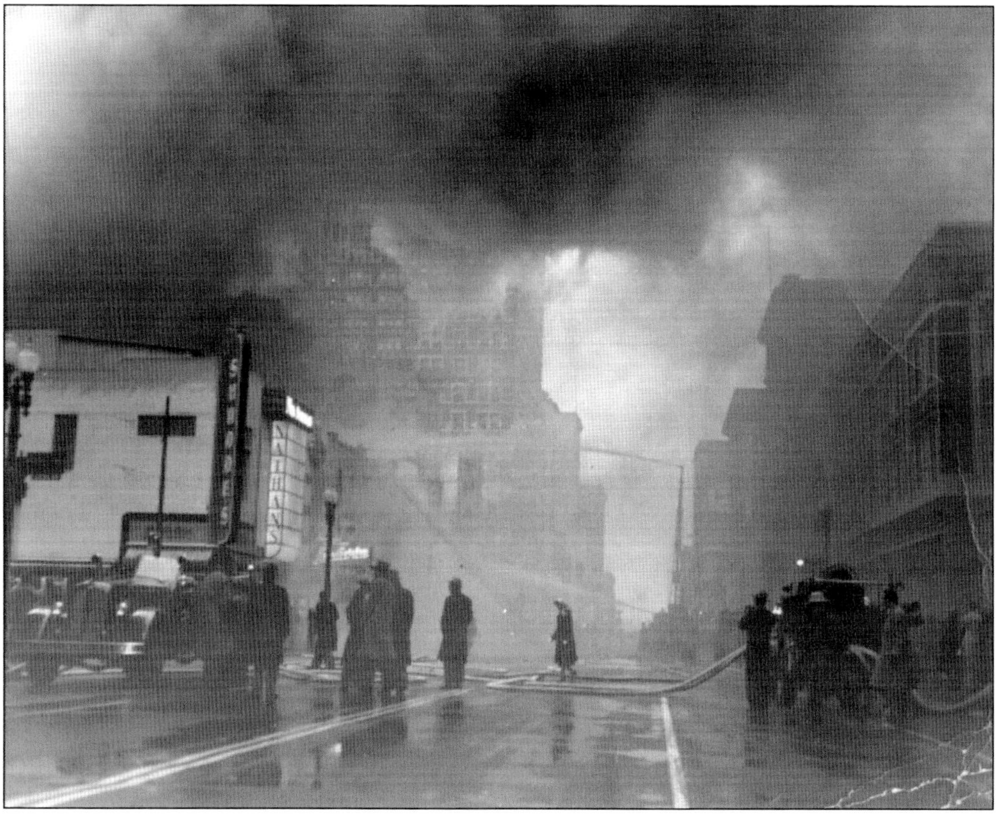

Assistant Fire Chief C.A. Middlekauf of Central Fire Station died of a heart attack at a March 10, 1953, fire on Center Street. He was directing the fire attack at Fedrick Poultry & Egg Company when he fell to his knees and asked for assistance. The chief was removed from the scene before he lost consciousness. Efforts to revive him were unsuccessful, and he was pronounced dead at the scene.

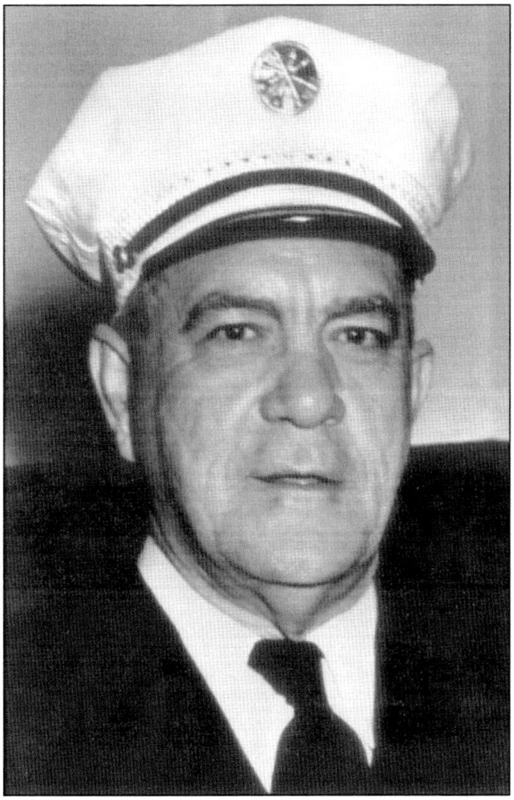

Pipe & Ladderman Fidel Chabolla of Station 1 (seen here in the 1950s) responded to a box alarm at Gable Street and the Missouri-Kansas-Texas (MKT) Railroad tracks on July 29, 1953. Ladder 1 was delayed leaving the station, and arrived at the intersection with Engine No. 8. The rigs collided and were demolished, killing Chabolla. Firefighter Joseph J. Solito of Engine No. 8 was also killed, and seven firefighters were injured.

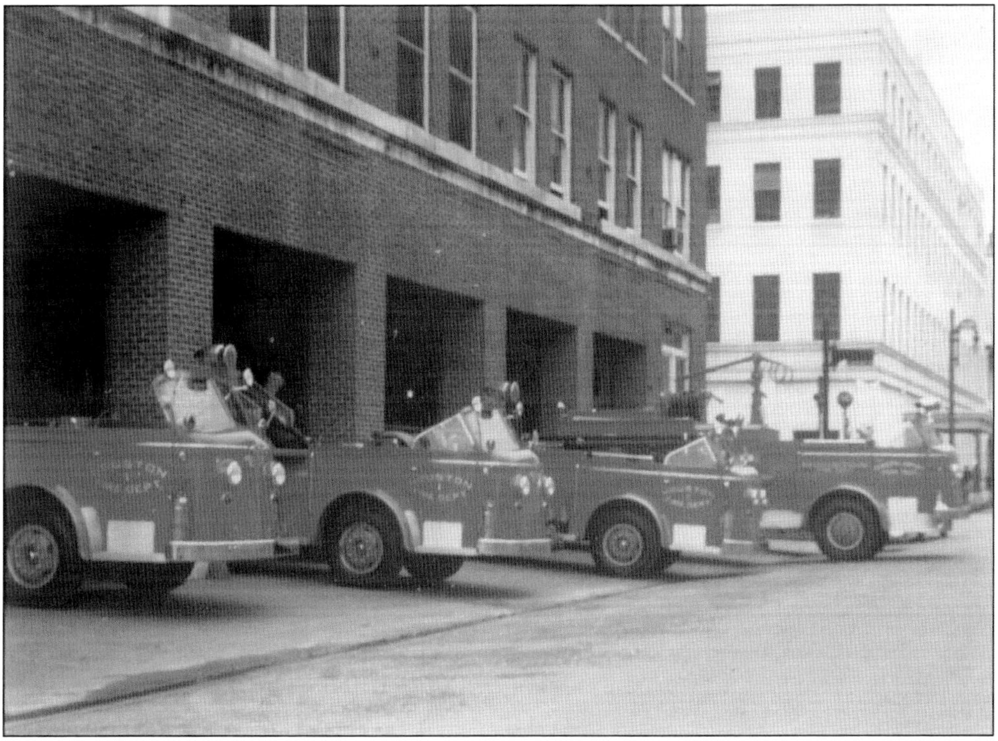

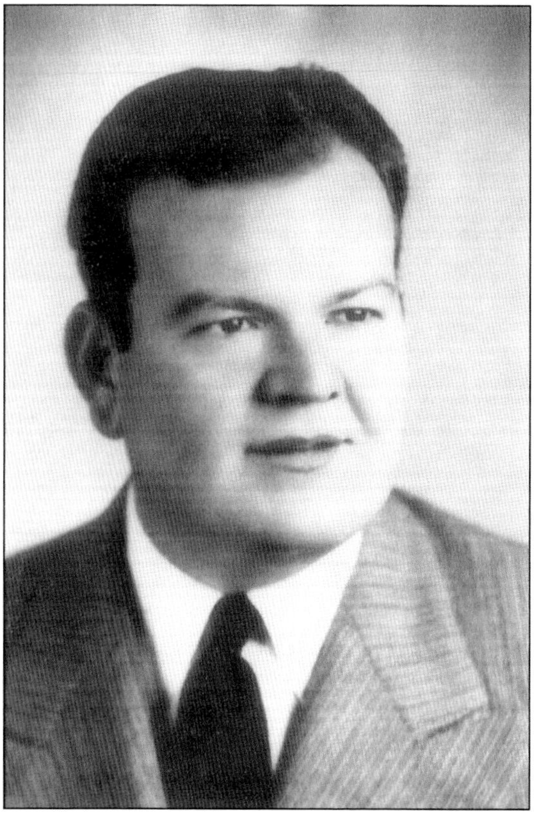

On the night of July 29, 1953, Chauffeur Joseph J. Solito was killed in a collision between Engine No. 8 and Truck No. 1 at the intersection of Preston and Crawford Streets. Solito was driving Engine 8 on the way to a box alarm at Gable Street and the MKT tracks. Another firefighter on Truck 1, Fidel Chabolla, was also killed in the accident. Solito had entered the department in 1938.

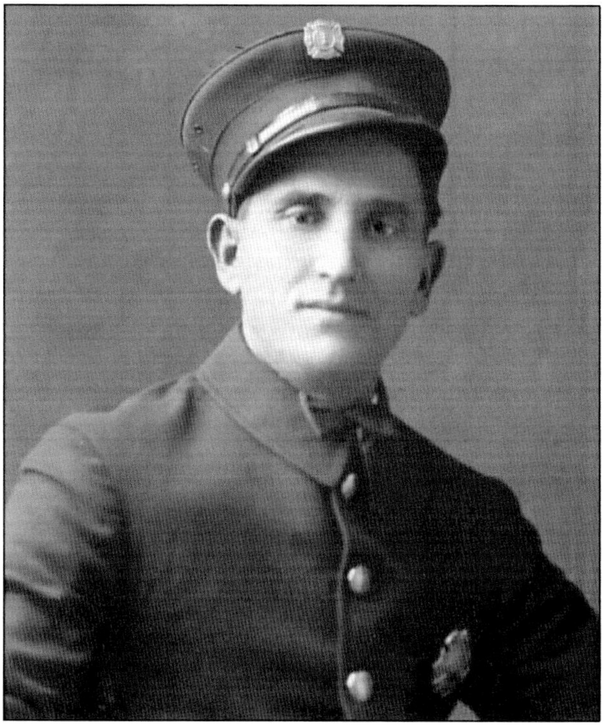

On January 7, 1954, Pipe & Ladderman Frank Catino of Central Fire Station was killed in an accident between Ladder No. 1 and an automobile at the intersection of Preston and Crawford Streets. The automobile sped through the intersection and was broadsided by Ladder 1. Catino, born Angelo Cortaneglia, entered the department on May 15, 1911. The fire chief renamed him because he could not pronounce "Cortaneglia"; the name stuck.

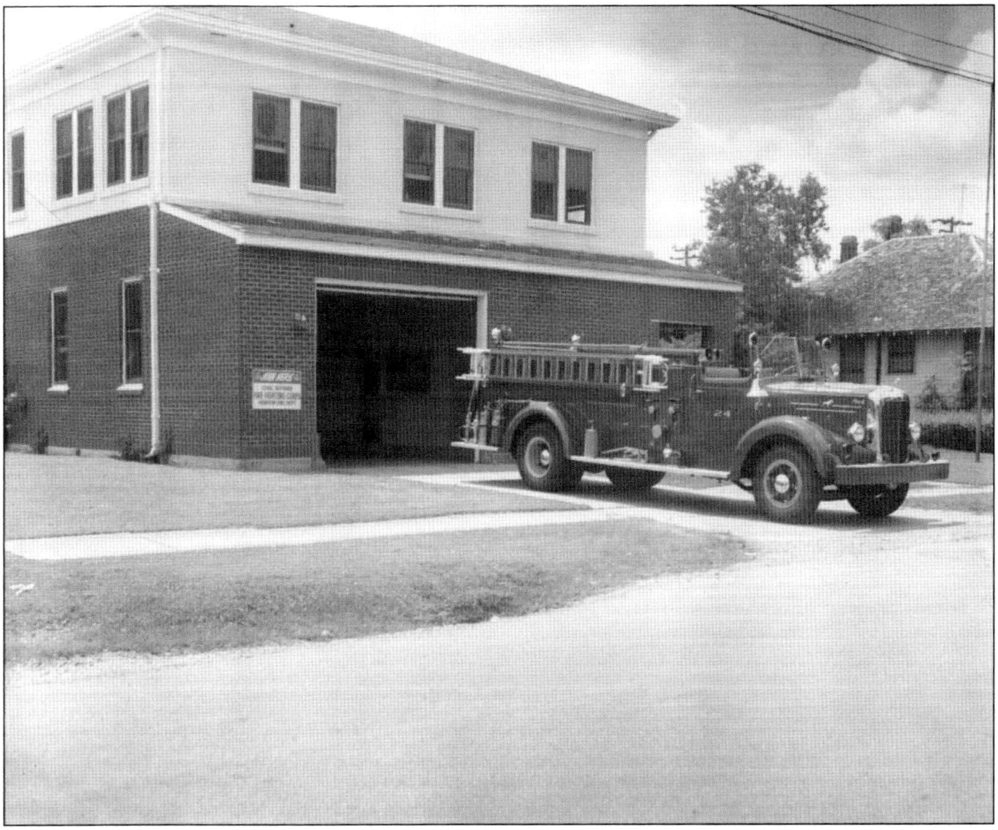

On February 23, 1954, Chauffeur Woodrow Erwin of Station 24 (pictured here) was acting captain filling in for the off-duty regular captain. The crew responded to a fire at a small shack at a junkyard located at 2705 Clay Street. Erwin complained of sharp chest pains and was transported to Jefferson Davis Hospital. He died shortly after arrival.

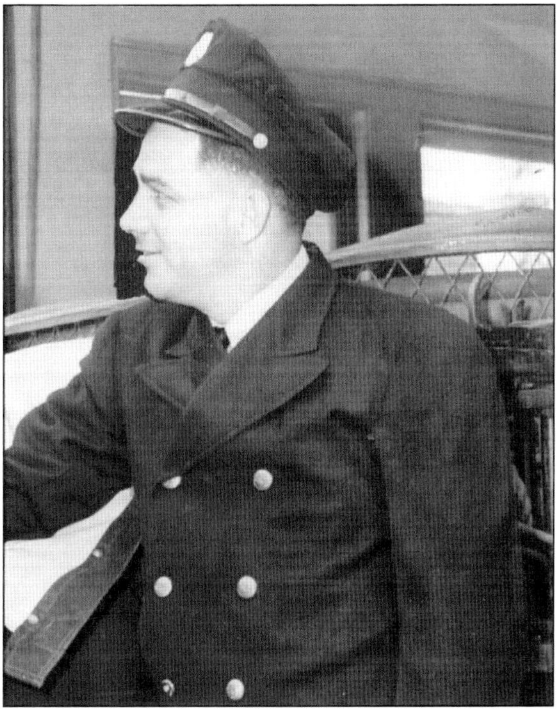

Chauffeur Delbert Sherfy, acting captain of Station 17, was killed in an accident on October 31, 1954. Sherfy's crew was responding to a washdown of fuel after an automobile collision. Sherfy's vehicle collided with another vehicle at Sampson and Canal Streets. He was killed instantly in the crash. In addition, four other firefighters and both occupants of the other car were injured, one of whom died the following day.

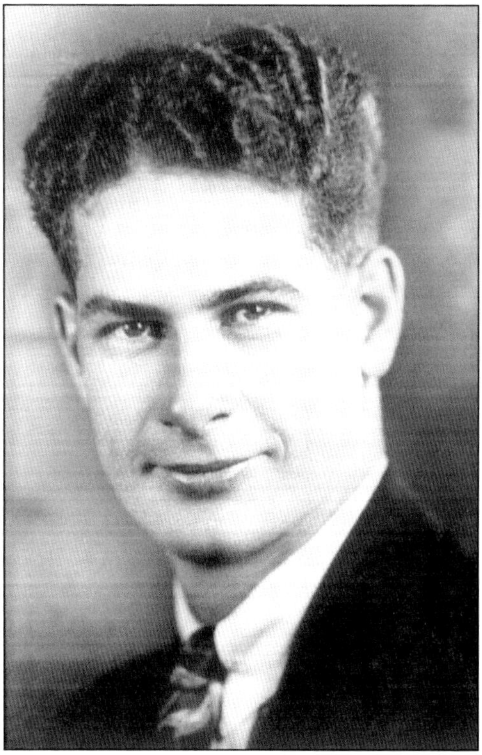

William Kersten, a chauffeur at Station 19, was killed on December 18, 1954, while en route to a fire at 7306 Bonham Street. He was driving the district chief when a city bus struck the car at the intersection of Lyons Avenue and Lockwood Drive. Kersten was transported to Jefferson Davis Hospital, where he died a short time later.

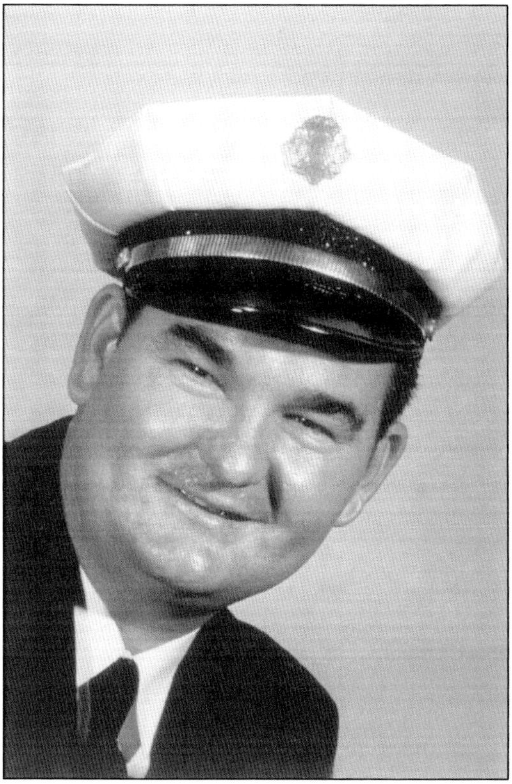

Pipe & Ladderman Willis McWhorter of Station 30 was killed at an accident on November 6, 1959, involving Engine No. 30 and a trailer truck loaded with steel. The accident occurred at the intersection of Fulton Street and Berry Road. All of the firefighters on Engine 30 were seriously injured; one of them, Charles Jedlicka, died 11 days later. The men were en route to a fire at Airline and Benbrook Drives.

Pipe & Ladderman Charles Jedlicka of Station 30 succumbed to injuries 11 days after an accident on November 6, 1959. Engine 30 was en route to a fire on Airline Drive when it was broadsided by a trailer truck loaded with steel at the intersection of Berry Road and Fulton Street. Jedlicka was flown to Brook Army Medical Center in San Antonio, where he died on November 17.

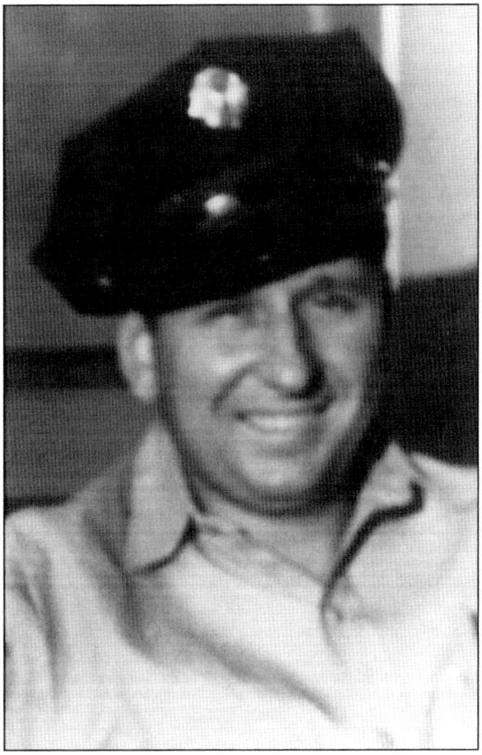

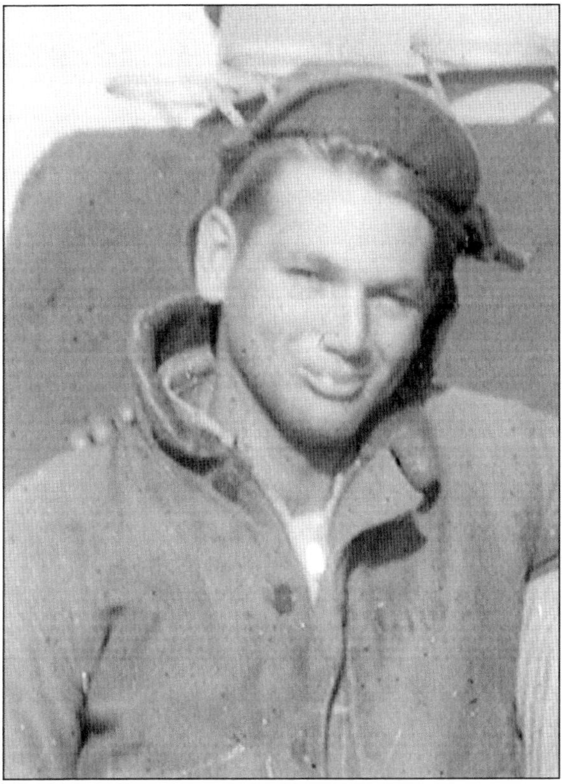

Pipe & Ladderman Doyle Chandler of Station 19 was killed at a fire aboard the SS *Amoco Virginia* on November 9, 1959. Chandler and other firefighters were searching for victims aboard the ship after a 16-hour battle to extinguish the blaze. In the darkness, Chandler fell through a hole in the deck and plunged into a partially filled hold of aviation gasoline. Chandler entered the department on July 1, 1958.

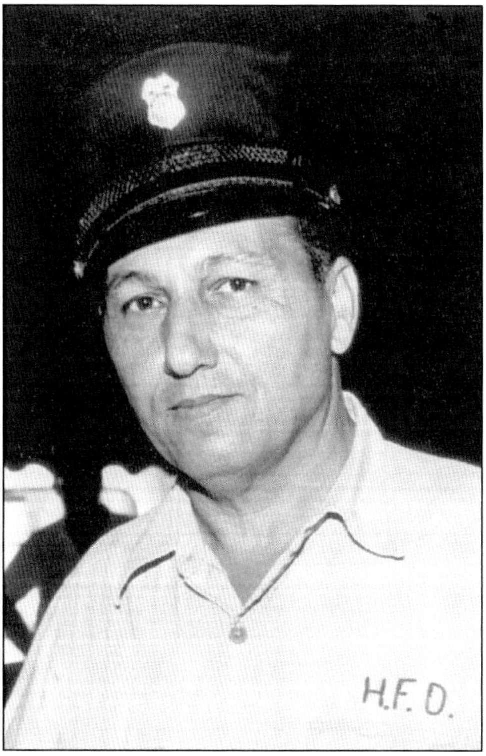

Don Contorno, a Pipe & Ladderman at Fire Station 37, suffered a heart attack on May 2, 1962, at a two-alarm fire involving Wanda Petroleum Company on Almeda Genoa Road. He was transported to St. Joseph Hospital and, after a 19-day struggle, died May 21. Contorno had entered the department on December 16, 1941.

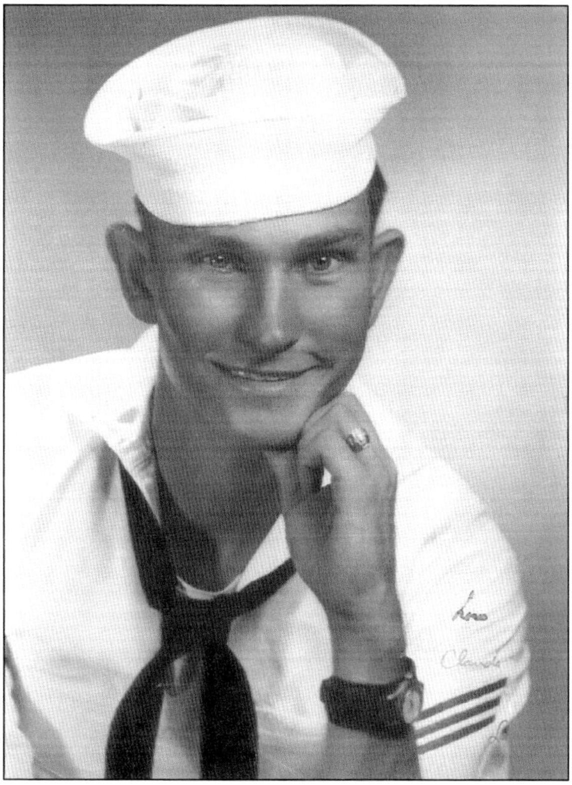

Chauffeur Claude Hopkins of Station 2 was killed on June 30, 1962, in a vehicular accident at Louisiana and Preston Streets. Someone at 1209 Commerce Street had attempted to get a drink from a sprinkler-system drain valve and set off the sprinkler alarm. Ladder No. 2, driven by Hopkins, responded and collided en route with a trailer truck loaded with grain.

Capt. Steve Baranowski of Station 37 suffered a fatal heart attack on February 3, 1969, while overhauling a house fire at 3631 Underwood Street. Overhaul was taking place to clean up and ensure the fire was extinguished. Captain Baranowski and his crew were on the roof, pulling shingles off. Baranowski suddenly fell unconscious and was brought down. Efforts to revive him at the scene failed.

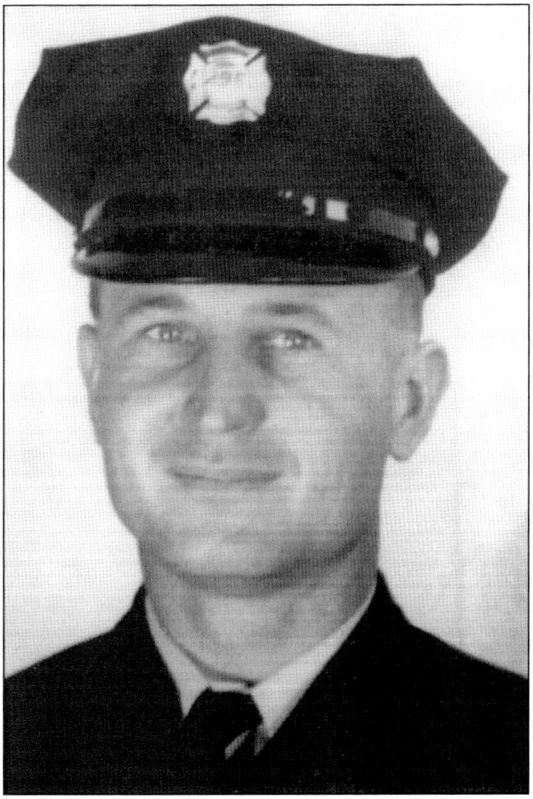

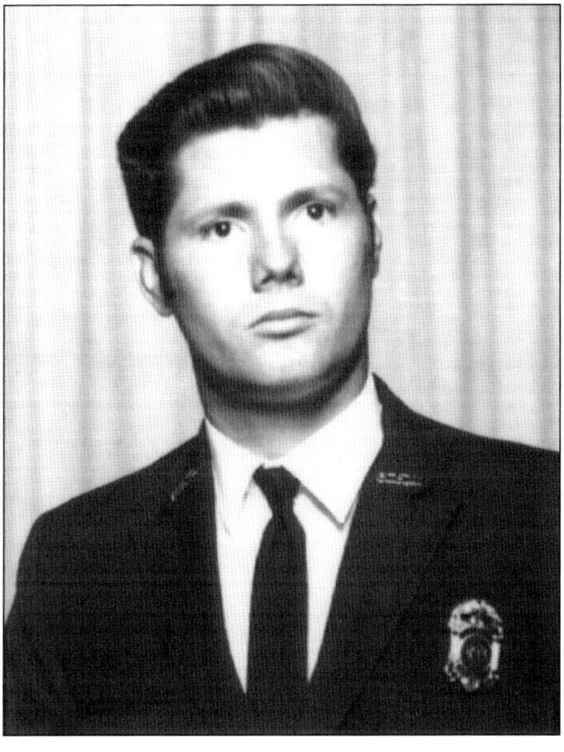

On May 3, 1971, Pipe & Ladderman James Walls of Station 51 was killed after his ambulance collided with another vehicle and overturned at the intersection of Chimney Rock and Westheimer Roads. The ambulance had a patient and EMT in the back, and both received only minor injuries. The accident occurred almost one month after the department began ambulance service with light-panel trucks.

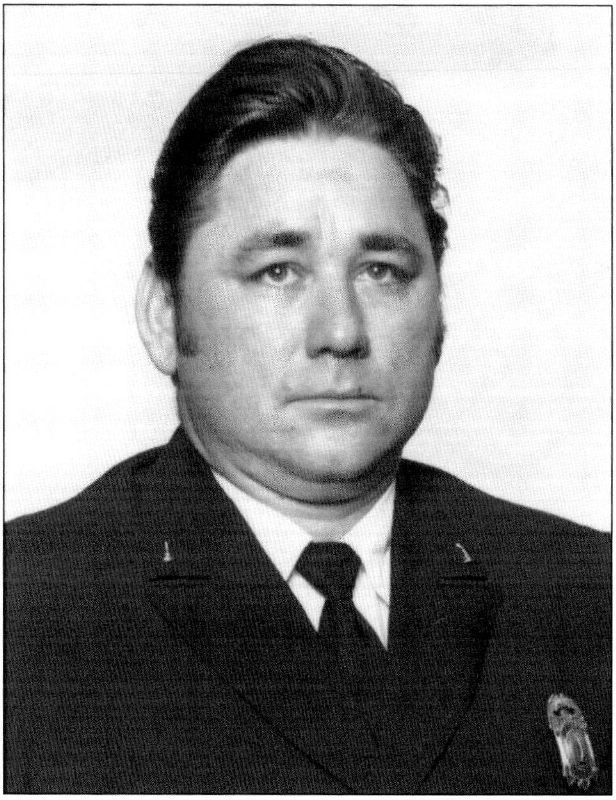

An explosion of a railroad tank car on Mykawa Road claimed the life of Fire Inspector Truxton Hathaway on October 19, 1971. One end of the separated tank car struck Hathaway as he was taking photographs of the fire. Hathaway had been temporarily assigned to the nearby Fire Training Academy. In addition, 27 other firefighters suffered burns at the 2-11, 6-tap fire. Hathaway had entered the department on February 15, 1961.

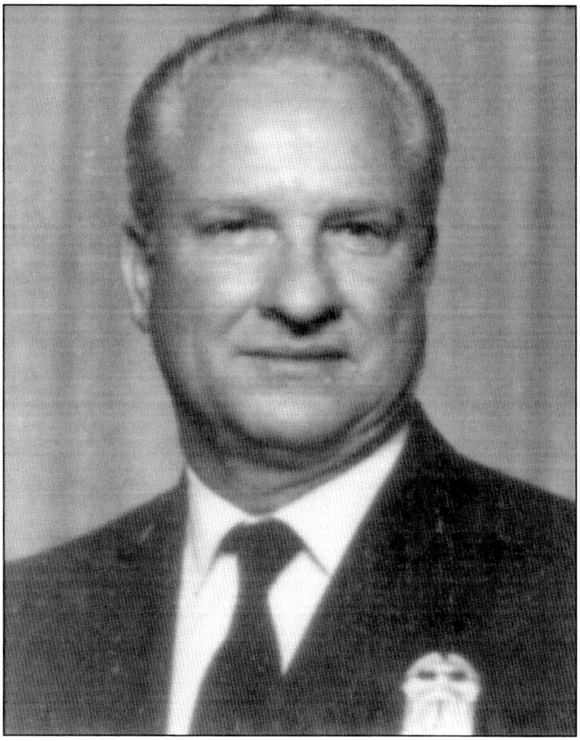

Lewis "Moe" Richardson, a captain at Station 7, died of a heart attack while battling a stubborn apartment fire on February 3, 1973. The blaze was at 1400 Richmond Avenue. Despite immediate efforts at the scene, Captain Richardson could not be revived. He had entered the fire department on August 1, 1941.

Sr. Capt. Jackie Ludwick, a 21-year HFD veteran, died of a heart attack on February 4, 1977. Ludwick, of Station 7, was acting district chief at a house fire that had been reported at Wheeler Avenue and Scott Street. He was transported to Hermann Hospital. Doctors attempted to save him, but all efforts were unsuccessful.

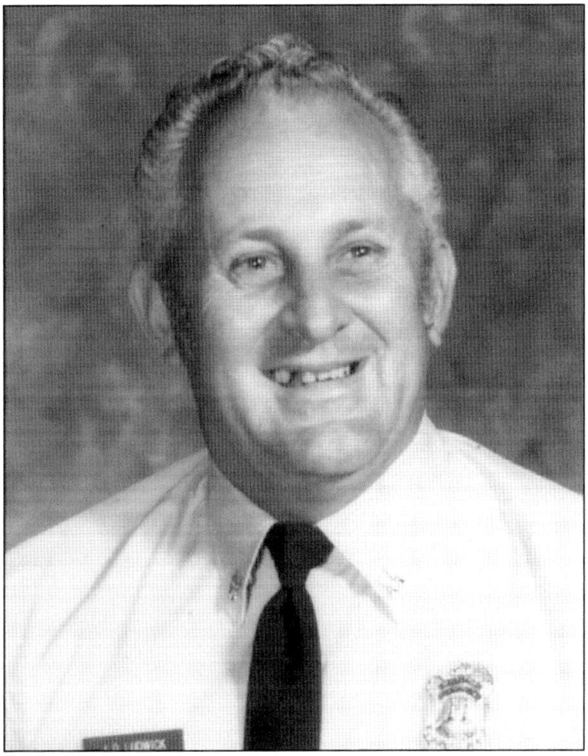

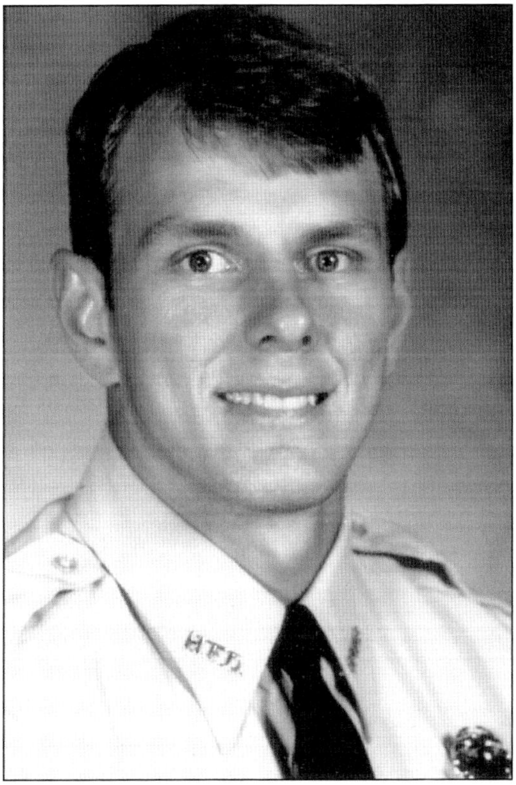

Probationary firefighter Thomas Cooper, assigned to Station 1, was injured on September 21, 1982, responding to a trash fire while out drilling. Hose was quickly rolled up and placed on the tailboard of the engine. En route to the fire, Cooper tried to retrieve a roll of hose that was falling as the engine rounded a corner, and he fell. Cooper died at St. Joseph Hospital two days later, on September 23.

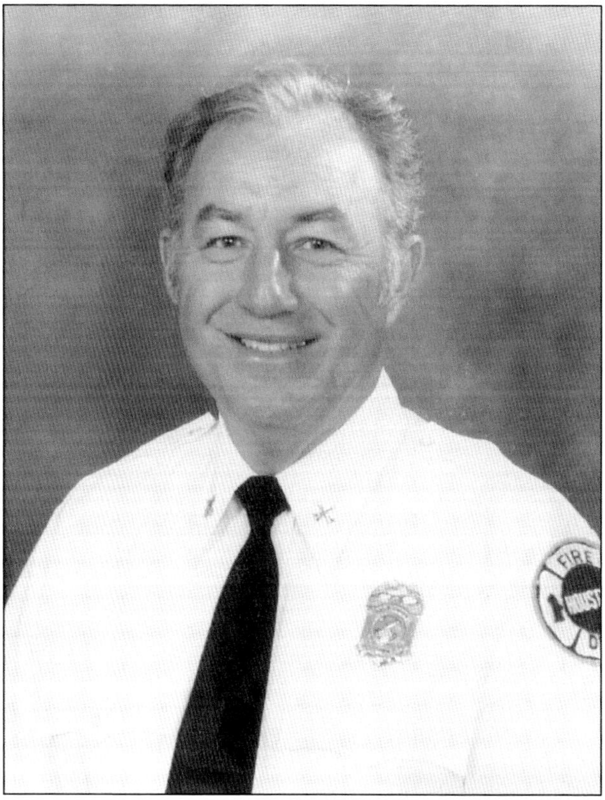

District Chief Lonnie Franklin of Station 7 was killed on January 4, 1983, in an accident at Elgin and Hutchins Street. He was responding to an early-morning alarm when an automobile ran a stop sign and slammed into the chief's car. Efforts to save his life at Hermann Hospital failed. Chief Franklin had entered the fire department on August 21, 1950.

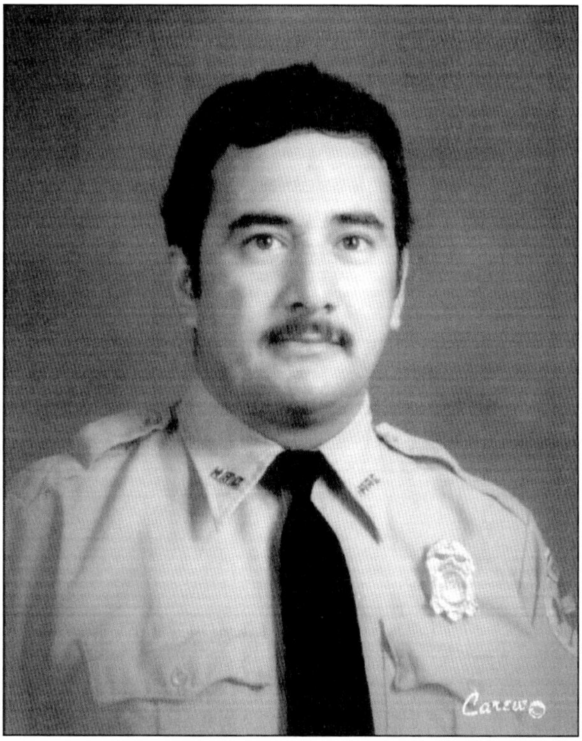

Pipe & Ladderman Robert Reyes of Station 62 was killed on December 1, 1982. Reyes was filling in on Engine No. 15 when he fell onto the North Freeway en route to a car fire. Reyes had spent 10 years with the Federal Bureau of Investigation before changing careers and entering the Houston Fire Department.

A flashover at a fire in a two-story residence on Roseneath Drive, near Station 25, killed Houston Volunteer Fire Department District Chief Ruben Lopez on December 4, 1996. Lopez was working out of Fire Station 25, where he usually rode his volunteer shifts. He was searching for an occupant when a flashover trapped him, eliminating escape. He died along with the occupant he was searching for.

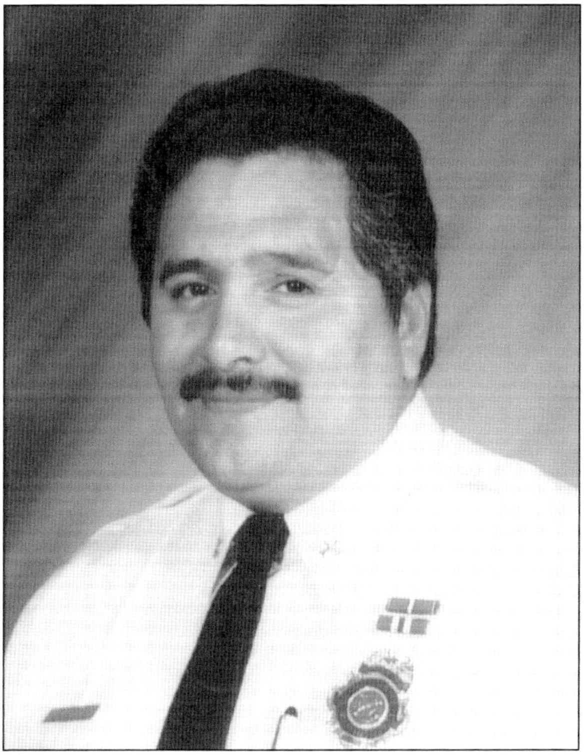

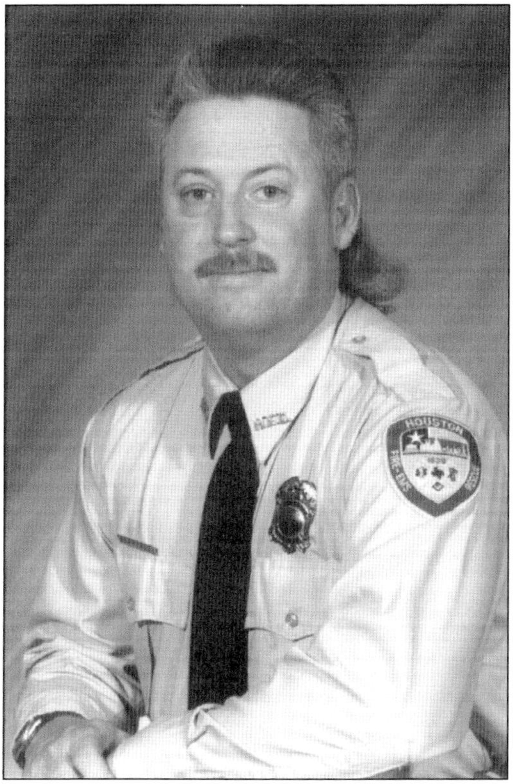

Firefighter Steven Mayfield died of a heart attack during a training exercise in Dallas on December 2, 1998. He was assigned to Airport Fire Station 92. The training exercise involved the extinguishment of an aircraft fire, one of the required training exercises for crash firefighters. Mayfield had entered the department on March 31, 1975.

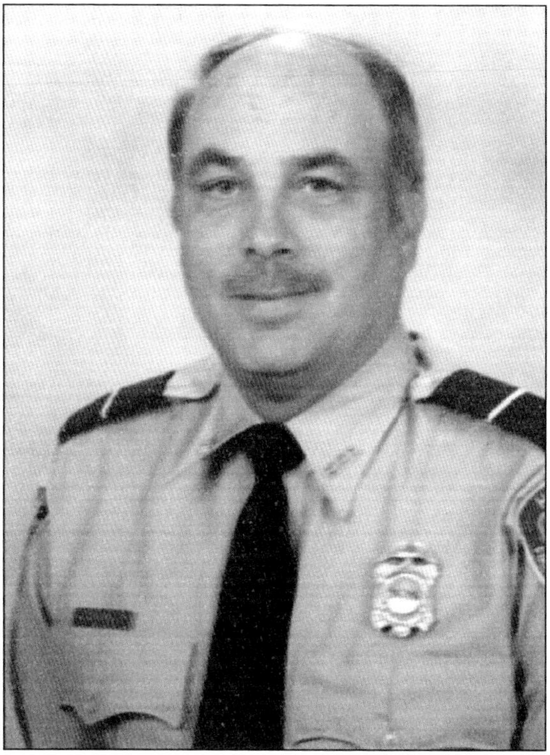

Firefighter Lewis Mayo (left) of Engine 76 was killed at an early-morning blaze in a McDonalds at 12602 Bissonnet Street on February 14, 2000. The roof collapsed shortly after the crew entered the burning structure. Falling debris separated Mayo and fellow firefighter Kimberly Smith (below, also of Engine 76) from their fire hose, a lifeline to the outside. Neither firefighter was able to locate the hose line; both perished.

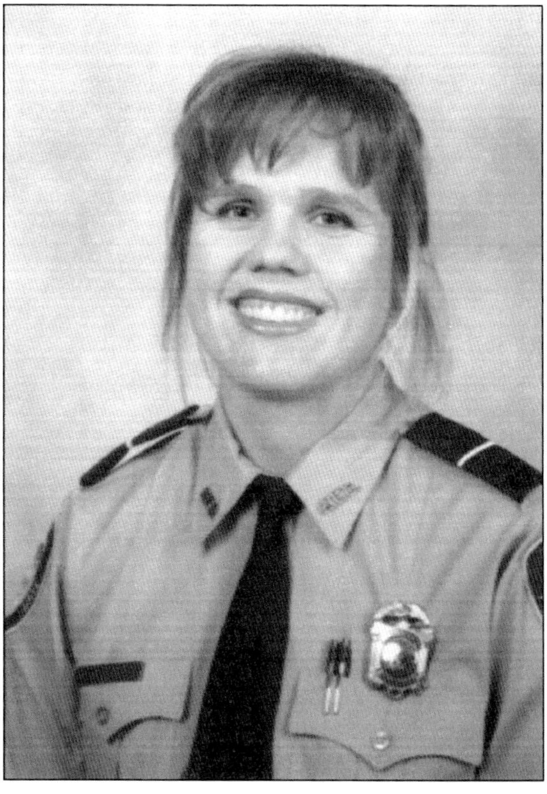

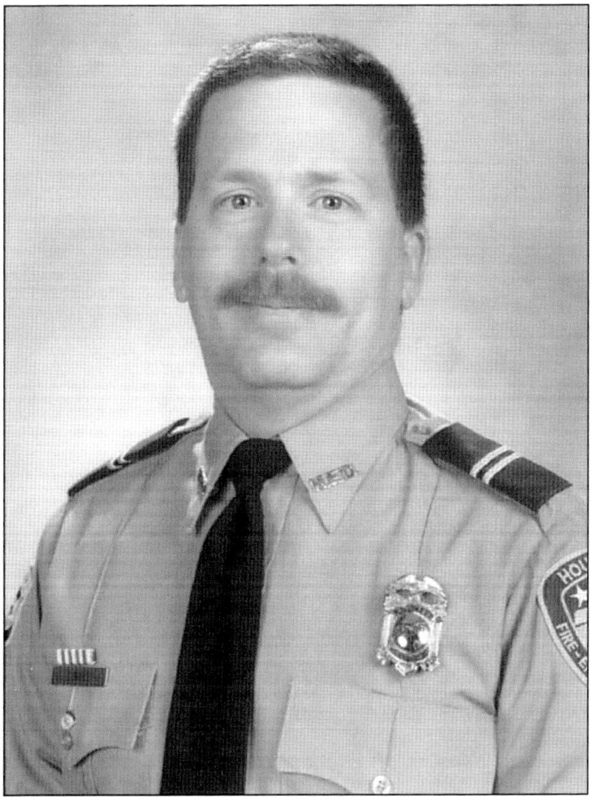

Capt. Grady Burke, with Station 46, responded to an early morning house fire on February 19, 2005, at 8510 Brandon Street. The house was abandoned, but it was being used by squatters. Captain Burke's crew was inside the structure, trying to extinguish the fire, when the roof collapsed. All of the crew members save Burke escaped. The man who set the fire was charged with and convicted of manslaughter.

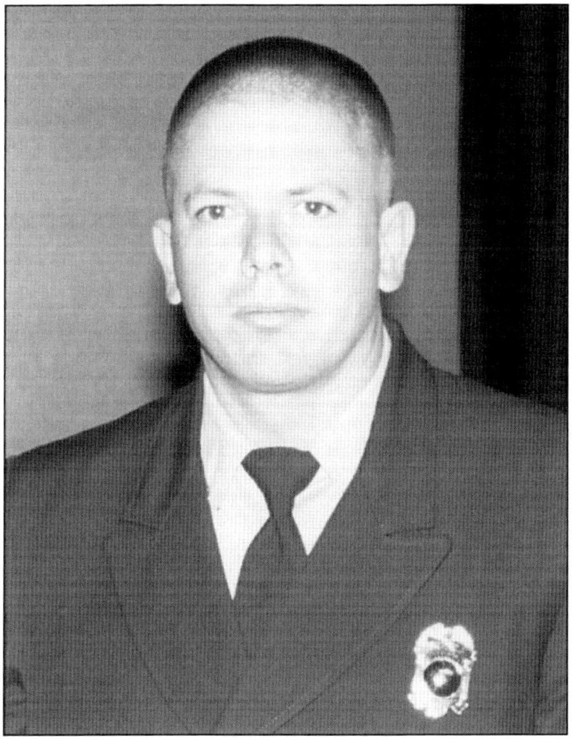

Firefighter Kevin Kulow of Station 50 was trapped when the roof of the El Festival ballroom on Kempwood Drive collapsed on April 4, 2004. Firefighters arrived to find the building filled with smoke, and bystanders told them that there were people inside. Shortly after the crew entered the building, the fire intensified, and the roof collapsed. Kulow was unable to reach the door, and died before being discovered.

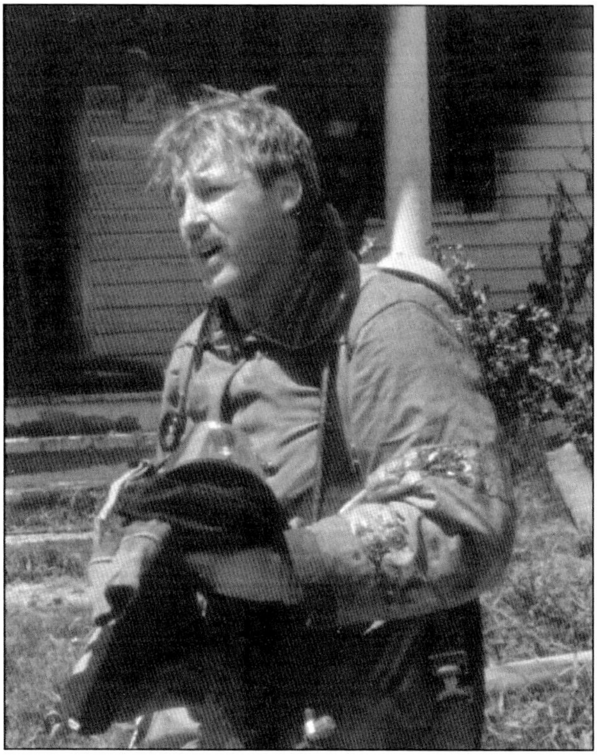

Firefighters responded to a six-alarm fire at a tower condominium building on San Felipe Street on October 13, 2001. Capt. Jay Jahnke, with Engine No. 2, proceeded to the fifth floor on a search-and-rescue mission. Heavy winds pushed fire and smoke on top of the firefighters. Jahnke became trapped, ran out of air, and was unable to make his escape. Efforts to revive him at the hospital were unsuccessful.

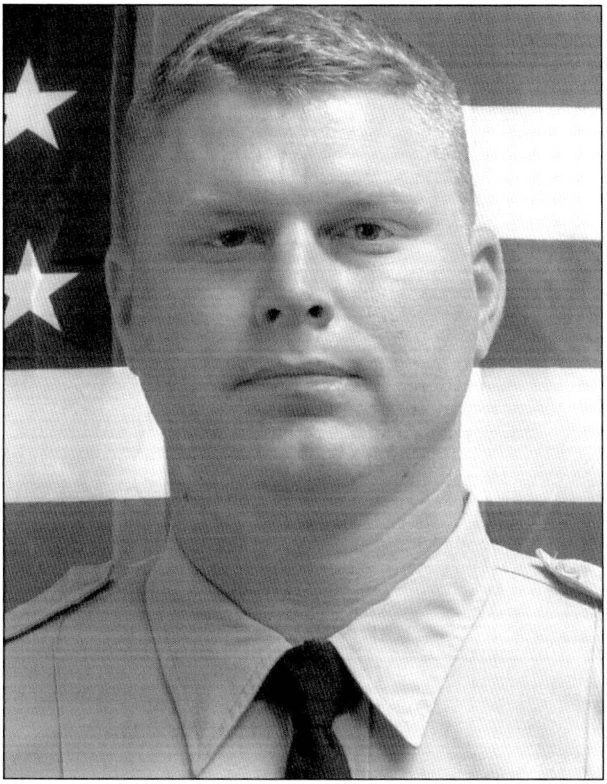

Damion Hobbs attained his dream job of firefighter after his military discharge following two tours in Iraq. Hobbs graduated from the training and was assigned to Station 26. During his first six months, he was scheduled to ride with Captain Harlow on Engine No. 26, then exhaust his final six months of probation on Ladder No. 26. The morning that he died would have been his last day to work on Engine 26.

On the early morning of April 12, 2009, Station 26 responded to a house fire on Oak Vista Street, just a couple miles from the station. Upon arrival, Capt. James Harlow (pictured) and rookie Damion Hobbs entered the home to begin the process of controlling the fire. Heavy winds blowing through the house rapidly worsened the conditions. The flames spread, causing the two firefighters to head unsuccessfully for the exit.

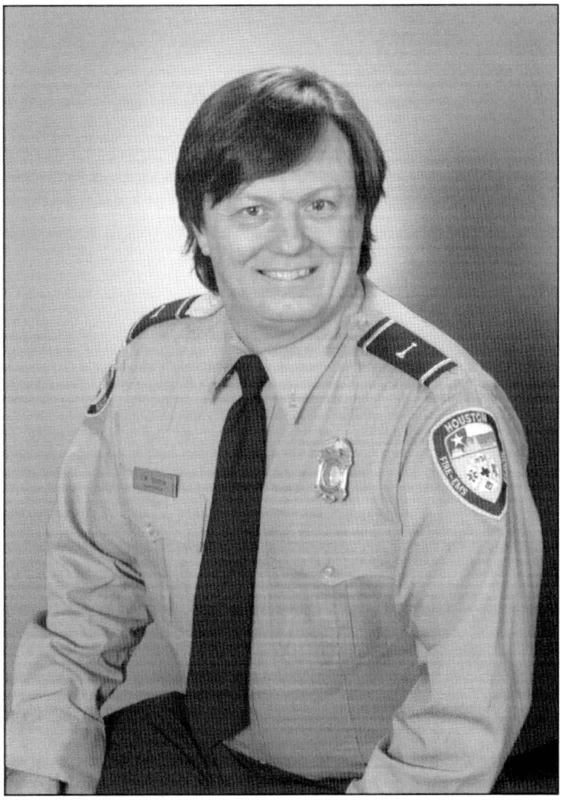

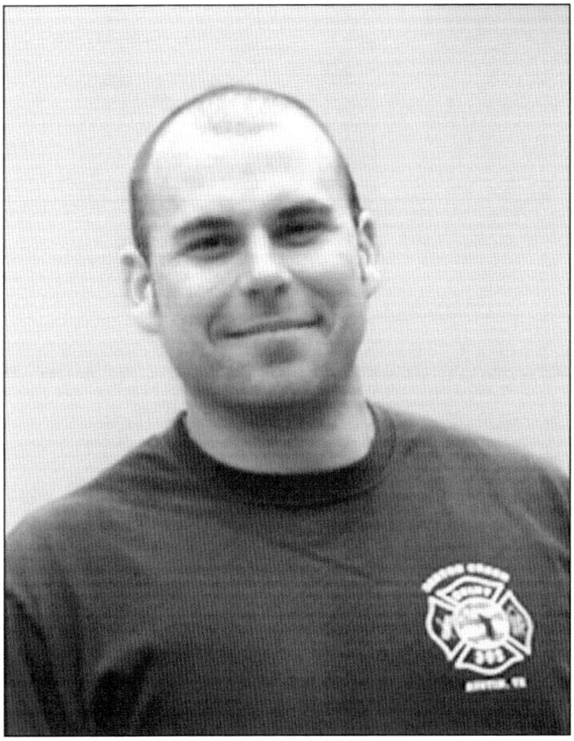

On April 29, 2009, HFD cadet Cohnway Johnson, just one month into his training, collapsed after completing his morning run. He was rushed to Memorial Hermann and admitted to the ICU, where he died five days later. Johnson had previously worked for the Yoakum Fire Department, east of San Antonio, and the Oak Hill Fire Department, south of Austin.

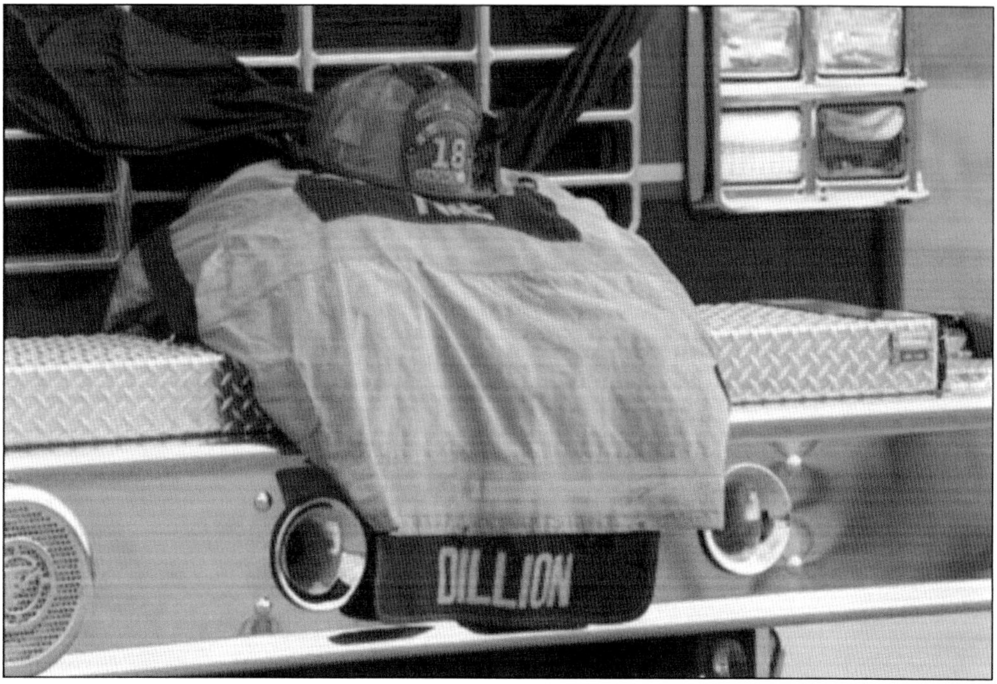

On March 14, 2012, Capt. Bill Dillion, usually out of Station 69, was working a debit day at Station 76. On the morning of his shift, Dillion's ladder company was dispatched to an apartment fire at 7520 Cook Road. The small kitchen fire was quickly extinguished, but the focus soon turned to the captain after he collapsed. He was rushed to the hospital, where a short time later he was pronounced dead.

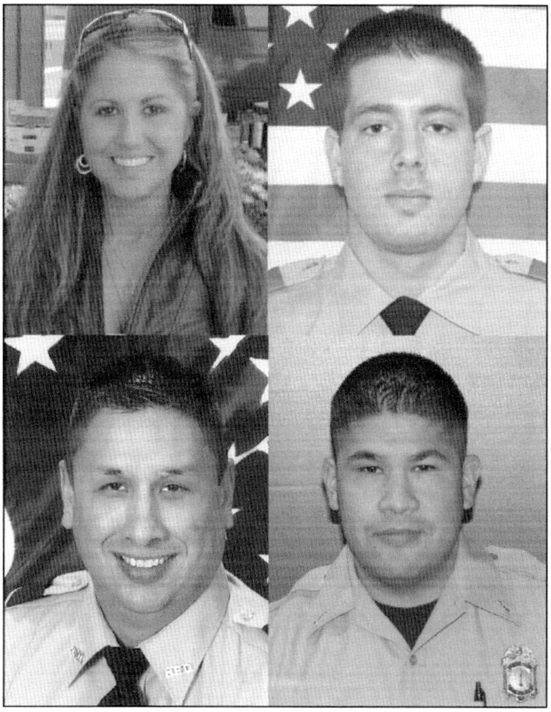

On Friday, May 13, 2013, a small restaurant fire turned into a massive blaze that engulfed the adjacent Southwest Inn on the Southwest Freeway. Part of the building collapsed as firefighters searched for people who might be trapped inside. Probationary firefighter Anne Sullivan and firefighter Robert Garner, of Station 68 (top row), and Engine Operator Robert Beebe (bottom left) and Capt. Matthew Renaud (bottom right), of Station 51, perished in the blaze.

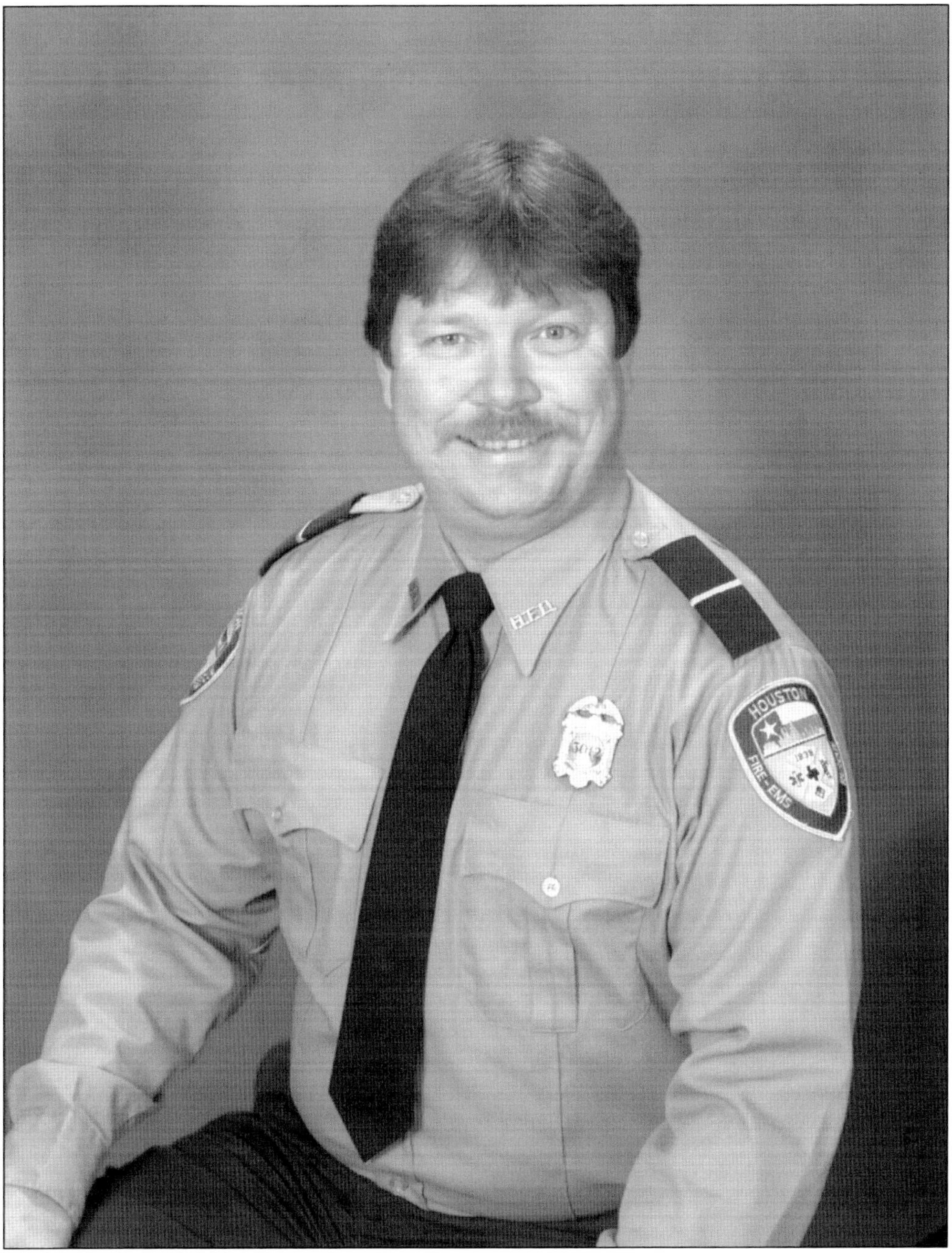

Daniel Groover, 46, was a 21-year veteran of the Houston Fire Department and the fifth firefighter to die in the line of duty in 13 months. At a two-alarm house fire, reported in the afternoon of July 9, 2014, in Kingwood, Groover collapsed while performing a search-and-rescue mission in the empty two-story house. He never recovered. Groover had served HFD as a paramedic and firefighter at Station 104.

Discover Thousands of Local History Books Featuring Millions of Vintage Images

Arcadia Publishing, the leading local history publisher in the United States, is committed to making history accessible and meaningful through publishing books that celebrate and preserve the heritage of America's people and places.

Find more books like this at
www.arcadiapublishing.com

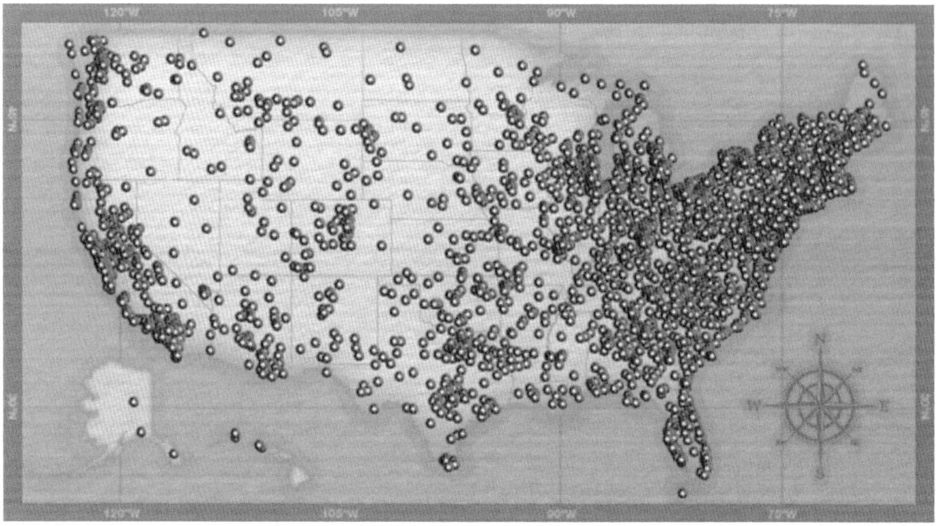

Search for your hometown history, your old stomping grounds, and even your favorite sports team.

Consistent with our mission to preserve history on a local level, this book was printed in South Carolina on American-made paper and manufactured entirely in the United States. Products carrying the accredited Forest Stewardship Council (FSC) label are printed on 100 percent FSC-certified paper.

MADE IN THE USA